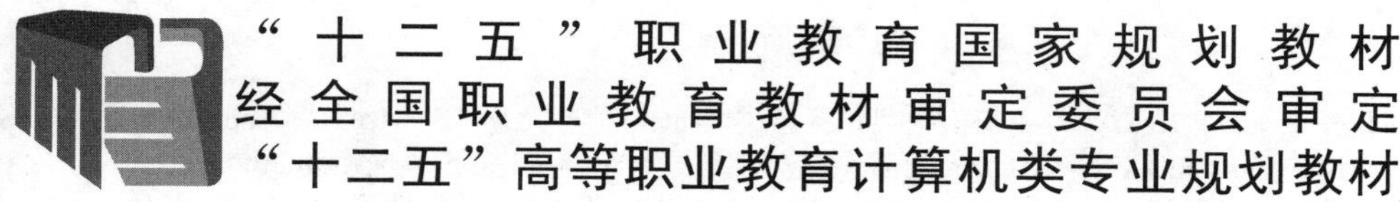

Android 高级应用编程实战

李华忠　梁永生　主编

刘立明　蔡　磊　叶炼炼　副主编

中国铁道出版社有限公司
CHINA RAILWAY PUBLISHING HOUSE CO., LTD.

内 容 简 介

全书共分 9 章，前 6 章为理论篇，主要包括意图（Intent）与服务（Service）、Android 数据永久存储应用、Android 网络应用、Android 调用外部数据、Android 多媒体应用和 Android 系统服务应用等核心理论知识；后 3 章为综合项目实训篇，主要包括三个综合实训项目（基于 Android 智能家居系统项目、Android 校园通项目和基于 Android 智能仓储系统项目），综合应用了本书介绍的核心知识和关键技术。

本书符合教学规律和课堂要求，很好地反映了嵌入式和移动互联等行业出现的 Android 方面的新知识、新技术、新方法和新应用，能解决高校 Android 课程教学面临的迫切问题，适合作为高等院校 Android 高级应用程序设计的教材，也可作为移动开发爱好者的自学参考书。

图书在版编目（CIP）数据

Android 高级应用编程实战 / 李华忠，梁永生主编. — 北京：中国铁道出版社，2015.1（2019.12重印）
“十二五”高等职业教育计算机类专业规划教材
ISBN 978-7-113-19061-3

Ⅰ. ①A… Ⅱ. ①李… ②梁… Ⅲ. ①移动终端－应用程序－程序设计－高等职业教育－教材 Ⅳ. ①TN929.53

中国版本图书馆 CIP 数据核字(2014)第 186763 号

书　　名：Android 高级应用编程实战
作　　者：李华忠　梁永生　主编

策　　划：王春霞　　　　**读者热线：**（010）63550836
责任编辑：王春霞　包　宁
封面设计：付　巍
封面制作：白　雪
责任校对：汤淑梅
责任印制：郭向伟

出版发行：中国铁道出版社有限公司（100054，北京市西城区右安门西街 8 号）
网　　址：http://www.tdpress.com/51eds/
印　　刷：北京鑫正大印刷有限公司
版　　次：2015 年 1 月第 1 版　　2019年12月第4次印刷
开　　本：787mm×1092mm　1/16　**印张：**16.5　**字数：**399 千
书　　号：ISBN 978-7-113-19061-3
定　　价：32.00 元

前言

Android是一个由谷歌（Google）和开放手机联盟（Open Handset Alliance）开发与倡导的，以Linux为基础的完整、开放、免费的手机平台。它由应用程序、应用程序框架、系统库、Android运行时以及Linux内核5部分组成。Android高级应用编程实战以面向对象Java语言实现的应用程序框架为基础，易学、易用，从而极大地降低了在手机和平板电脑等终端设备上开发移动互联应用程序的难度，大大提高了Apps应用程序开发的效率，已经成为世界上主流移动应用程序开发平台之一。

目前，我国很多院校的计算机软件、移动互联、嵌入式和物联网等相关专业，都将“Android高级应用编程实战”作为一门专业核心课程。为了帮助院校老师能够比较全面、系统地讲授这门课程，使学生能够熟练地使用Android高级技术进行移动互联软件开发，我们几位长期在院校从事Android教学的教师和企业工程师，共同编写了这本《Android高级应用编程实战》教材。

我们对本书的体系结构做了精心的设计，按照Android平台的技术体系结构和项目内容，如意图（Intent）与服务（Service）、Android数据永久存储、Android网络、Android调用外部数据、Android多媒体和Android系统服务等项目，设计多个学习情境。每个学习情境又结合知识体系和实践技能细化为若干个子学习情境，由浅入深，实用性强。最后，结合移动互联应用实际情况，安排三个综合实训项目（基于Android智能家居系统、Android校园通和基于Android智能仓储系统），在提高学生应用技能的同时，强化项目驱动，实施“工学结合”，提高理论教学和实践教学质量，充分满足了高职院校对教学和学生自学的需求。在内容编写方面，我们注意难点分散、循序渐进；在文字叙述方面，我们注意言简意赅、重点突出；在实例选取方面，我们注意实用性强、针对性强。

本书理论篇每章都附有一定数量的习题，可以帮助学生进一步巩固基础知识；本书综合项目实训篇每章还附有实践性较强的项目实施，可以供学生上机操作时使用。本书配备了PPT课件、源代码、习题答案、教学大纲、课程设计等丰富的教学资源，任课教师可到中国铁道出版社网站（http://www.51eds.com/）免费下载使用。本书的教学参考总学时为64学时，其中实践环节为20学时，各章的参考学时参见下面的学时分配表。

章　节	课程内容	学时分配	
		讲　授	实　训
第1章	意图（Intent）与服务（Service）	4	2
第2章	Android数据永久存储应用	4	2
第3章	Android网络应用	6	2
第4章	Android调用外部数据	6	2
第5章	Android多媒体应用	6	2
第6章	Android系统服务应用	6	2

续表

章　节	课程内容	学时分配	
		讲　授	实　训
第7章	基于Android智能家居系统项目	6	4
第8章	Android校园通项目	6	4
第9章	基于Android智能仓储系统项目(可选)		
课时总计		44	20

本书由深圳信息职业技术学院李华忠副教授和梁永生教授担任主编；深圳尚信物联有限公司研发总监刘立明、深圳市优科赛服网络科技有限公司研发总监蔡磊和厦门海洋职业技术学院叶炼炼担任副主编共同完成；深圳信息职业技术学院软件学院院长邓果丽、副院长但唐仁、嵌入式技术与应用专业郑洪英、吴险峰、曾路和盛建强、软件技术专业覃国蓉、唐强平和软件测试专业的房丽娜、陈勖、何涛等同事对本书的编写提出了很多宝贵意见，我们在此表示衷心的感谢！尤其感谢深圳市大雅新科技有限公司为本书提供全套智能家居系统设备和项目源代码。

由于移动开发技术发展日新月异，加之我们水平有限，书中难免存在疏漏和不妥之处，敬请广大读者批评指正。

编　者

2014年11月

CONTENTS

目 录

理 论 篇

第 1 章 意图（Intent）与服务（Service）......2

1.1 学习导入......2
1.1.1 什么是意图......2
1.1.2 什么是服务......2
1.1.3 Android 平台应用开发技术回顾（Activity 等四大组件技术）......3
1.2 技术准备......3
1.2.1 意图......3
1.2.2 服务......9
1.3 案例......18
1.3.1 Android 应用程序过渡页面......18
1.3.2 服务器/客户端通信中的心跳包功能......20
1.4 知识扩展......24
1.4.1 BroadcastReceiver（广播接收器）......24
1.4.2 数据绑定 Bundle 的主要功能函数......24
1.4.3 意图的主要功能函数......24
1.5 本章小结......25
1.6 强化练习......25

第 2 章 Android 数据永久存储应用......26

2.1 学习导入......26
2.1.1 什么是 SharedPreferences（偏好数据存储）......26
2.1.2 什么是文件存储数据......26
2.1.3 什么是 SQLite 数据库存储数据......27
2.1.4 什么是 ContentProvider 存储数据......27
2.1.5 什么是网络存储数据......27
2.2 技术准备......27
2.2.1 SharedPreferences 存储数据......27
2.2.2 文件存储数据......29
2.2.3 SQLite 数据库存储数据......35
2.2.4 ContentProvider 存储数据......43
2.2.5 网络存储数据......49
2.3 案例......50
2.3.1 基于 sharedPreferences 的参数配置文件......50
2.3.2 基于 SQLite 数据库的设备状态信息显示......55
2.4 知识扩展......60

2.5 本章小结 60
2.6 强化练习 60

第 3 章 Android 网络应用 61

3.1 学习导入 61
3.1.1 窄带广域网 61
3.1.2 宽带广域网 62
3.1.3 HTTP 通信 63
3.1.4 Socket 通信 63
3.1.5 Wi-Fi 63
3.1.6 蓝牙 67
3.2 技术准备 73
3.2.1 Android 网络基础 73
3.2.2 HTTP 通信 75
3.2.3 Socket 通信 81
3.3 案例 85
3.3.1 迷你浏览器 85
3.3.2 获取 Web 网址信息 86
3.4 知识扩展 90
3.4.1 使用 WebView 浏览网页 90
3.4.2 使用 WebView 中 JavaScript 脚本调用 Android 方法 90
3.5 本章小结 90
3.6 强化练习 90

第 4 章 Android 调用外部数据 91

4.1 学习导入 91
4.2 技术准备 91
4.2.1 SAX 解析 xml 文件 91
4.2.2 DOM 解析 xml 文件 93
4.2.3 基于位置的服务 95
4.2.4 基于地图的应用 96
4.3 案例——Web 服务中的 Json 数据解析 102
4.4 知识扩展 108
4.4.1 根据 GPS 信息在地图上定位 108
4.4.2 调用 Google 的地址解析服务 108
4.5 本章小结 109
4.6 强化练习 109

第 5 章 Android 多媒体应用 110

5.1 学习导入 110
5.2 技术准备 110
5.2.1 使用 MediaPlayer 播放音频 110
5.2.2 使用 SoundPool 播放音频 120
5.2.3 使用 VideoView 播放视频 122

5.2.4 使用 MediaPlayer 与 SurfaceView 播放视频 123
5.2.5 使用 MediaRecorder 录制音频 126
5.2.6 使用摄像头拍照 129
5.2.7 控制摄像头录制视频短片 134
5.3 案例——流媒体播放器 137
5.4 知识扩展 140
5.4.1 传感器知识 140
5.4.2 传感器的典型案例 140
5.5 本章小结 140
5.6 强化练习 140

第 6 章 Android 系统服务应用 142

6.1 学习导入 142
6.2 技术准备 143
6.2.1 活动管理器（ActivityManager） 143
6.2.2 警报管理器（AlarmManager） 144
6.2.3 音频管理器（AudioManager） 148
6.2.4 剪贴板管理器（ClipboardManager） 156
6.2.5 通知管理器（NotificationManager） 156
6.3 案例——网络诊断案例 158
6.4 知识扩展 171
6.4.1 电话管理器（TelephonyManager） 171
6.4.2 短信管理器（SmsManager） 171
6.5 本章小结 171
6.6 强化练习 171

综合项目实训篇

第 7 章 基于 Android 智能家居系统项目 174

7.1 项目概述 174
7.1.1 平台简介 174
7.1.2 硬件资源 175
7.2 项目设计 176
7.2.1 项目总体功能需求 176
7.2.2 项目总体设计 177
7.3 必备的技术和知识点 178
7.4 项目实施 179
7.4.1 登录页面 179
7.4.2 主页面 180
7.4.3 情景模式页面 183
7.4.4 我的房间页面 187
7.4.5 具体设备页面 192
7.4.6 家用电器页面 195

7.5 本章小结 200

第 8 章 Android 校园通项目 201

8.1 项目概述 201
8.2 项目设计 201
8.2.1 项目总体功能需求 201
8.2.2 项目总体设计 202
8.3 必备的技术和知识点 209
8.3.1 DAO 设计模式 209
8.3.2 XML 解析 209
8.3.3 服务器数据访问 212
8.3.4 异步任务 AsyncTask 213
8.4 项目实施 214
8.4.1 导入项目所需要的资源 214
8.4.2 建立数据库相关的包 215
8.4.3 添加 DAO 类和 Java 对象类 218
8.4.4 自定义列表视图，实现下拉刷新功能 219
8.4.5 定义列表适配器类，实现列表展现 220
8.4.6 修改主页面的布局和 Activity 类，完成列表展示 223
8.4.7 上传数据库文件 224
8.4.8 添加详细信息页面 225
8.4.9 从服务器更新数据 228
8.4.10 搭建服务器环境进行测试 230
8.5 本章小结 232

第 9 章 基于 Android 智能仓储系统项目 233

9.1 项目概述 233
9.2 项目设计 233
9.2.1 项目总体功能需求 233
9.2.2 项目总体设计 234
9.3 必备的技术和知识点 235
9.4 项目实施 235
9.4.1 登录页面 235
9.4.2 主页面 237
9.4.3 环境监控页面 240
9.4.4 物品入库页面 245
9.4.5 具体设备页面 248
9.4.6 物品出库页面 252
9.5 本章小结 255

参考文献 256

理论篇

本篇主要系统讲解 Android 高级应用编程项目中涉及的具有共性、可重用性和普遍性的一些核心理论知识。以项目引导的创新思路，每章首先明确各共性知识的学习目标，通过学习导入，循序渐进地引导学生，做必要的技术准备，然后通过典型案例讲解技术实施步骤和关键实现代码，最后，通过知识扩展和强化练习，夯实学生的理论和实践基本素养和技能。

本篇主要包含以下 6 章：

第 1 章　意图（Intent）与服务（Service）
第 2 章　Android 数据永久存储应用
第 3 章　Android 网络应用
第 4 章　Android 调用外部数据
第 5 章　Android 多媒体应用
第 6 章　Android 系统服务应用

意图（Intent）与服务（Service）

学习目标：

- 理解 Intent 对于 Android 应用的作用；
- 理解 Intent 元素的组成及其作用；
- 理解意图过滤器（IntentFilter）的作用；
- 理解 Service 对于 Android 应用的作用；
- 理解 Service 的生命周期；
- 掌握使用 Intent 启动系统组件的方法；
- 掌握 Service 启动到销毁过程的实现；
- 掌握 Service 与 Activity 通信的实现。

1.1 学习导入

1.1.1 什么是意图

意图（Intent）是一种运行时绑定（Runtime Binding）机制，它能在程序运行时动态地连接两个不同的组件。即 Intent 提供了一种通用的消息系统，它允许在应用程序与其他应用程序间传递 Intent 来执行动作和产生事件。使用 Intent 可以激活 Android 应用的三个核心组件：活动（Activity）、服务（Service）和广播接收器（BroadcastReceiver）。

（1）通过 startActivity( ) 或 startActivityForResult( )启动一个 Activity；

（2）通过 startService( ) 启动一个服务，或者通过 bindService( ) 和后台服务交互；

（3）通过广播方法（sendBroadcast(), sendOrderedBroadcast(), sendStickyBroadcast()）发给广播接收器（BroadcastReceiver）。

1.1.2 什么是服务

服务（Service）是运行在后台的线程，级别与 Activity 差不多。既然说 Service 是运行在后台的线程，那么它就是不可见的，没有界面。用户可以启动一个 Service 播放音乐，或者记录地理信息位置的改变，或者启动一个服务一直监听某种动作。

Service 和其他组件一样，都运行在主线程中，因此不能利用它做耗时的请求或者动作。用户可以在服务中开一个线程，在线程中做耗时动作。

1.1.3 Android 平台应用开发技术回顾（Activity 等四大组件技术）

Android 四大基本组件分别是活动（Activity）、服务（Service）、内容提供者（Content Provider）和广播接收器（BroadcastReceiver）。

（1）在应用程序中，一个 Activity 通常就是一个单独的屏幕，它既可以显示一些控件，也可以监听并处理用户的事件并作出响应。Activity 之间通过 Intent 进行通信。

（2）一个 Service 是一段长生命周期的，没有用户界面的程序，可以用来开发如监控类程序。

（3）Content Provider 将应用程序的指定数据集提供给其他应用程序。这些数据可以存储在文件系统、SQLite 数据库或任何其他合理的方式中，其他应用程序可以通过 ContentResolver 类从该内容提供者中获取或存入数据。

（4）BroadcastReceiver 是一种被广泛运用在应用程序之间传输信息的机制。通过 BroadcastReceiver 对发送出来的广播进行过滤接收并响应一类组件。

1.2 技术准备

1.2.1 意图

1. **显式 Intent**

显式 Intent 是指在调用中通过代码直接指定哪个组件应该处理该 Intent，但是由于开发者往往并不清楚应用程序组件的名称，因此显式 Intent 大多用于在应用程序内部传递消息。在调用过程中通过指定接收器的 Class 或 ComponentName 实现。要显示地调用一个 Intent，可以使用 Intent.setClass（Context context，Class class）形式，通过这种方法可以直接传递 Activity 或 Service 类的引用处理 Intent，从而简化整个 Android 的 Intent 解析过程。

下面的示例程序示范了如何通过显式 Intent 启动另一个 Activity。

```
Intent intent=new Intent();     //构造意图对象
intent.setClass(MainActivity.this, SecondActivity.class);
                                //显示方式声明 Intent，直接启动 SecondActivity
startActivity(intent);          //启动 Activity
```

2. **隐式 Intent**

隐式 Intent 是指在调用过程中由 Android 系统决定哪个组件应该处理该 Intent。在实现过程中通过一个 Intent 解析过程实现，该解析过程使用了动作（Action）、数据（Data）和类别（Category）。由于隐式 Intent 没有明确的目标组件名称，所以要由 Android 寻找与 Intent 请求意图最匹配的组件，具体方法是：Android 将 Intent 的请求内容和一个称为 IntentFilter 的过滤器进行比较，IntentFilter 中包含系统所有可能的待选组件。如果 IntentFilter 中某一组件匹配隐式 Intent 请求的内容，那么 Android 就选择该组件作为该隐式 Intent 的目标组件。

3. **意图过滤器（IntentFilter）**

在显示 Intent 和隐式 Intent 中，已经讲述了 IntentFilter 的作用是告诉 Android 系统，应用程序本身所能处理和响应的隐式 Intent 请求的内容。需要在 Android 应用程序的系统控制文件 AndroidManifest.xml 中添加<intent-filter>标签对 IntentFilter 进行声明。

每个<intent-filter>元素被解析成一个 IntentFilter 对象。当将一个 apk 的包安装到 Android 系统中时，其中的组件就会向平台注册。一旦 Android 系统建立了一个 IntentFilter 的注册表项，系统就会知道如何把收到的 Intent 请求映射到已注册的 Activity、BraodcastReceiver 和 Service。

请求 Intent 时，Android 平台使用 Intent 的 Action、Data 和 Category 作为标准，通过已经注册的 IntentFilter 开始解析过程。Intent 和 IntentFilter 匹配的原则如下：

（1）动作和类别必须匹配。

（2）如果指定数据类型的话，则数据类型必须匹配，或者数据方案、授权和路径的组合必须匹配。

一个隐式 Intent 请求要想顺利传递到目标组件，必须通过 Action、Data 和 Category 这三方面的检查，若三方面都不匹配，那么系统将不会把该隐式 Intent 传递给目标组件。

4. Intent 元素

Intent 元素包括：Action（动作）、Data（数据）、Category（类别）、Type（数据类型）、component（组件）和 extras（附加信息）。

（1）Action（动作）：用来指明要实施的动作是什么，比如 ACTION_VIEW、ACTION_EDIT 等。详细使用信息可查阅 Android SDK-> reference 中的 Android.content.intent 类，里面的 constants 中定义了所有的 Action。

（2）Data（数据）：要实施的具体的数据，一般由一个 Uri 变量表示。

下面是一些简单的例子：

```
ACTION_VIEW  content://contacts/1  //显示 identifier 为 1 的联系人的信息
ACTION_DIAL  content://contacts/1  //给这个联系人打电话
```

（3）Category（类别）：指定将要执行的 Action 的其他一些额外信息，例如 LAUNCHER_CATEGORY 表示 Intent 的接受者应该在 Launcher 中作为顶级应用出现；而 ALTERNATIVE_CATEGORY 表示当前的 Intent 是一系列可选动作中的一个，这些动作可以在同一块数据上执行。详细使用方法可参考 Android SDK-> reference 中的 Android.content.intent 类。

（4）Type（数据类型）：显式地指定 Intent 的数据类型（MIME）。一般 Intent 的数据类型能够根据数据本身进行判定，但是通过设置这个属性，可以强制采用显式指定的类型而不再进行推导。

（5）component（组件）：指定 Intent 的目标组件的类名称。通常 Android 会根据 Intent 中包含的其他属性信息（如 Action、Data/Type、Category）进行查找，最终找到一个与之匹配的目标组件。但是，如果指定 component 属性的话，将直接使用它指定的组件，而不再执行上述查找过程。指定了这个属性以后，Intent 的其他所有属性都是可选的。

（6）extras（附加信息）：是其他所有附加信息的集合。使用 extras 可以为组件提供扩展信息，例如，如果要执行“发送电子邮件”这个动作，可以将电子邮件的标题、正文等保存在 extras 里，传给电子邮件发送组件。

下面是这些额外属性的两个例子：

```
ACTION_MAIN with category CATEGORY_HOME       //用来启动主屏幕
ACTION_GET_CONTENT with MIME type vnd.android.cursor.item/phone
                                              //列出所有人的电话号码
```

5. Intent 典型应用案例

（1）Activity 之间的互相切换，详细代码参见配套光盘中的 CH01_1。

这个程序将会在程序中提供两个按钮，分别位于两个 Activity 中，通过这两个按钮实现两个 Activity 之间的互相切换。该程序的界面布局如图 1-1 和图 1-2 所示。

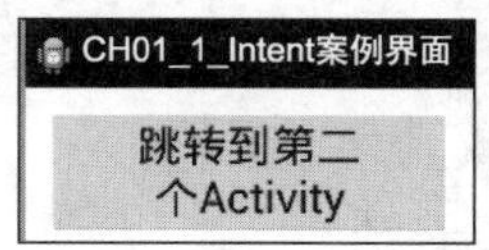

图 1-1　Intent 主界面

图 1-2　Intent 次界面

该程序的界面布局代码如下：CH01\CH01_1\res\layout\ activity_main.xml。

```
<RelativeLayout xmlns:android="http://schemas.android.com/apk/res/android"
    xmlns:tools="http://schemas.android.com/tools"
    android:layout_width="match_parent"
    android:layout_height="match_parent"
    android:paddingBottom="@dimen/activity_vertical_margin"
    android:paddingLeft="@dimen/activity_horizontal_margin"
    android:paddingRight="@dimen/activity_horizontal_margin"
    android:paddingTop="@dimen/activity_vertical_margin"
    tools:context=".MainActivity" >
    <Button
        android:id="@+id/mainButton"
        android:layout_width="wrap_content"
        android:layout_height="wrap_content"
        android:layout_alignParentTop="true"
        android:layout_centerHorizontal="true"
        android:text="@string/goto_second"
        android:textSize="32sp" />
</RelativeLayout>
```

程序清单：CH01\CH01_1\res\layout\ second.xml。

```
<LinearLayout xmlns:android="http://schemas.android.com/apk/res/android"
   android:layout_width="fill_parent"
   android:layout_height="fill_parent"
   android:orientation="vertical" >
   <TextView
       android:id="@+id/ShowData"
       android:layout_width="fill_parent"
       android:layout_height="wrap_content"
       android:textSize="40sp" />
   <Button
       android:id="@+id/returnButton"
       android:layout_width="match_parent"
       android:layout_height="wrap_content"
       android:text="@string/return_pre"
       android:textSize="32sp" />
</LinearLayout>
```

该程序的 Java 代码如下：CH01\CH01_1\src\com\zigcloud\ch01_1\ MainActivity.java。

```
public class MainActivity extends Activity {
                                //从 Activity 基类派生子类 MainActivity
    private Button mainButton=null;                     //定义按钮对象
    protected void onCreate(Bundle savedInstanceState) {   //重写子类方法
        super.onCreate(savedInstanceState);             //调用基类方法
        setContentView(R.layout.activity_main);         //设置主活动界面布局
        mainButton=(Button)findViewById(R.id.mainButton);
                                //根据按键 id 查找在布局文件中定义的按钮
        mainButton.setOnClickListener(new mainButtonListener());
                                                        //为按钮绑定事件监听器
    }
    class mainButtonListener implements OnClickListener{  //实现单击监听器
        public void onClick(View v) {                   //处理单击回调事件
            Intent intent=new Intent();                 //构造意图对象 intent
            //显示方式声明 Intent，直接启动 SecondActivity
            intent.setClass(MainActivity.this, SecondActivity.class);
            //从 MainActivity 向 SecondActivity 传递数据

            intent.putExtra("LihzMessage",getResources().getString(R.string.
bundle_string));
            startActivity(intent);                      //启动 Activity
        }
    }
}
```

程序清单：CH01\CH01_1\src\com\zigcloud\ch01_1\ SecondActivity.java。

```
public class SecondActivity extends Activity {   //从 Activity 基类派生子类
    TextView myText = null;                     //定义文本视图对象
    private Button returnButton=null;           //定义按钮对象
    protected void onCreate(Bundle savedInstanceState) {
                                                //重写子类 onCreate 方法
        super.onCreate(savedInstanceState); //调用基类 onCreate 方法
        setContentView(R.layout.second);        //设置第二个活动布局界面
        returnButton=(Button)findViewById(R.id.returnButton);
                                                //根据 id 从第二个布局中查找按钮对象
        returnButton.setOnClickListener(new SecondButtonListener());
                                                //设置按键单击监听事件
        myText=(TextView)findViewById(R.id.ShowData);
                                  //根据 id 从第二个布局中查找文本视图对象
        Intent intent=getIntent();              //获得启动该活动的意图对象
        String data=intent.getStringExtra("LihzMessage");
                                  //从意图对象中根据 LihzMessage 取得信息
        myText.setText(data);                   //将该信息显示在文本视图控件上
    }
    class SecondButtonListener implements OnClickListener{//实现单击监听事件
        public void onClick(View v) {           //处理单击回调方法
            Intent intent=new Intent();         //构造意图对象
            intent.setClass(SecondActivity.this, MainActivity.class);
                                                //从第二个活动向主活动切换
```

```
            startActivity(intent);                    //启动新活动
        }
    }
}
```

该程序运行效果如图 1-3 和图 1-4 所示。

（2）查看并获取联系人电话，具体代码请见配套光盘中的 CH01_2。

该程序提供一个按钮，用户单击该按钮时会显示系统的联系人列表，当用户单击指定联系人时，程序将会显示该联系人的名字和电话信息。该程序的界面布局如图 1-5 所示。

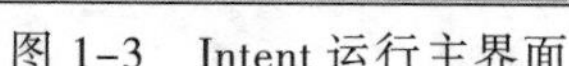

图 1-3　Intent 运行主界面

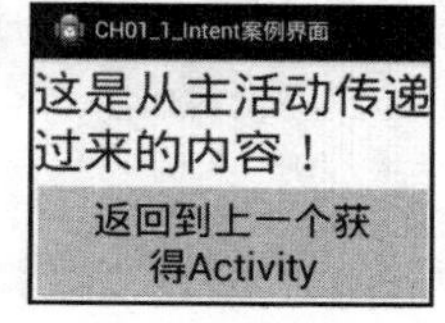

图 1-4　Intent 运行次界面

图 1-5　系统应用主界面

该程序的界面布局代码如下：CH01\CH01_2\res\layout\ activity_main.xml。

```
<?xml version="1.0" encoding="utf-8"?>
<LinearLayout xmlns:android="http://schemas.android.com/apk/res/android"
    android:orientation="vertical"
    android:layout_width="fill_parent"
    android:layout_height="fill_parent"
    android:gravity="center_horizontal">
<!-- 显示联系人姓名的文本框 -->
<EditText
    android:id="@+id/show"
    android:layout_width="fill_parent"
    android:layout_height="wrap_content"
    android:editable="false"
    android:cursorVisible="false"/>
<!-- 显示联系人电话的文本框 -->
<EditText
    android:id="@+id/phone"
    android:layout_width="fill_parent"
    android:layout_height="wrap_content"
    android:editable="false"
    android:cursorVisible="false"/>
<Button
    android:id="@+id/bn"
    android:layout_width="wrap_content"
    android:layout_height="wrap_content"
    android:text="查看联系人"
    android:textSize="32sp"   />
</LinearLayout>
```

程序的 Java 代码如下：CH01\CH01_2\ src\com\zigcloud\action\ MainActivity.java。

```
public class MainActivity extends Activity
{
    final int PICK_CONTACT=0;
    public void onCreate(Bundle savedInstanceState)
```

```
    {
        super.onCreate(savedInstanceState);
        setContentView(R.layout.activity_main);
        Button bn=(Button) findViewById(R.id.bn);
        bn.setOnClickListener(new OnClickListener() // 为bn按钮绑定事件监听器
        {
            public void onClick(View arg0)
            {
                Intent intent=new Intent();    //创建Intent
                //设置Intent的Action属性
                intent.setAction(Intent.ACTION_GET_CONTENT);
                intent.setType("vnd.android.cursor.item/phone");
                //设置Intent的Type属性
                //启动Activity，并希望获取该Activity的结果
                startActivityForResult(intent, PICK_CONTACT);
            }
        });
    }
    public void onActivityResult(int requestCode,int resultCode,Intent data)
    {
        super.onActivityResult(requestCode, resultCode, data);
        switch(requestCode)
        {
            case(PICK_CONTACT):
                if(resultCode==Activity.RESULT_OK)
                {
                    Uri contactData=data.getData(); // 获取返回的数据
                    CursorLoader cursorLoader=new CursorLoader(this,
contactData, null, null, null, null);
                    Cursor cursor=cursorLoader.loadInBackground();
                                                    // 查询联系人信息
                    if(cursor.moveToFirst())        // 如果查询到指定的联系人
                    {
                        String contactId=cursor.getString(cursor
.getColumnIndex(ContactsContract.Contacts._ID));
                        // 获取联系人的名字
                        String name=cursor.getString(cursor.getColumnIndexOrThrow(
ContactsContract.Contacts.DISPLAY_NAME));
                        String phoneNumber="此联系人暂未输入电话号码";
                        //根据联系人查询该联系人的详细信息
                        Cursor phones=getContentResolver().query(
ContactsContract.CommonDataKinds.Phone.CONTENT_URI, null,
ContactsContract.CommonDataKinds.Phone.CONTACT_ID+"="+contactId,null,null);
                        if(phones.moveToFirst())
                        { //取出电话号码
                            phoneNumber=phones.getString(phones
.getColumnIndex(ContactsContract.CommonDataKinds.Phone.NUMBER));
                        }
                        phones.close();                //关闭游标
                        EditText show=(EditText) findViewById(R.id.show);
```

```
                show.setText(name);            //显示联系人的名称
                EditText phone=(EditText) findViewById(R.id.phone);
                phone.setText(phoneNumber);  //显示联系人的电话号码
            }
            cursor.close();                    // 关闭游标
        }
        break;
    }
  }
}
```

该程序的运行效果如图 1-6 所示。

图 1-6 系统应用运行主界面

1.2.2 服务

1. Service 的两种模式

本地服务（Local Service）和远程服务（Remote Service）。

本地服务用于应用程序内部；远程服务用于 Android 系统内部的应用程序之间。前者用于实现应用程序自己的一些耗时任务，比如查询升级信息，并不占用应用程序（如 Activity 所属线程），而是单独开启线程后台执行，这样用户体验比较好。

2. Service 的生命周期

Android Service 的生命周期并不像 Activity 那么复杂，它只继承了 onCreate()、onStart()和 onDestroy()三个方法，当第一次启动 Service 时，先后调用 onCreate()和 onStart()两个方法，当停止 Service 时，则执行 onDestroy()方法，这里需要注意的是，如果 Service 已经启动了，当再次启动 Service 时，不会再执行 onCreate()方法，而是直接执行 onStart()方法。

- 启动服务：Context.startService(Intent service,Bundle b)
- 绑定服务：Context.bindService(Intent service,ServiceConnection c,int flag)

启动服务相当于告诉 Android 平台，在后台启动该 Service 并保持其运行，该服务与任何其他 Activity 或应用程序没有任何特定的连接。

绑定服务是指获得远程对象的句柄并从 Activity 中调用已定义的方法。因为每个 Android 应用程序都在其自己的进程中运行，所以使用已绑定的 Service 可以在不同进程之间传递数据。

（1）被启动服务的生命周期。如果一个 Service 被 Context.startService()方法启动，那么不管是否有活动绑定到该 Service，该 Service 都在后台运行。在这种情况下，如果需要该 Service 的 onCreate()方法将被调用，然后是 onStart()方法。如果一个 Service 被启动不止一次，则 onStart()方法将被多次调用，但是不会创建该 Service 的其他实例。该 Service 将一直在后台运行，直到被 Context.stopService()方法或其自己的 stopSelf()方法显示地停止该服务为止。

（2）被绑定服务的生命周期。如果一个 Service 被某个 Activity 调用 Context.bindService()方法绑定，则只要该连接被建立，则该服务将一直运行。Activity 可以使用 Context 建立到服务的连接，同时也要负责关闭该连接。当 Service 只是以这种方式被绑定而未被启动时，其 onCreate()方法被调用，但是 onStart()方法不被调用。在这种情况下，当绑定被解除时，平台可以停止和清除该 Service。

（3）被启动又被绑定服务的生命周期。如果一个 Service 既被启动又被绑定，则该 Service 基本上将在后台保持运行，与被启动服务的生命周期类似。唯一的区别在于生命周期本身。因为既启动服务又绑定服务，所以 onStart()和 onCreate()两个方法都将被调用。

（4）当服务被停止时清除服务。当一个 Service 被停止时，onDestory()方法或者在服务被启动以后被显式地调用，或者当不再有绑定的连接（没有被启动）时被隐式地调用。在 onDestory()方法中，每个 Service 都应该执行最终的清除功能，如停止任何已生成的进程等。

（5）Service 与 Activity 通信。通常每个应用程序都在它自己的进程内运行，但有时需要在进程间传递对象，可以通过应用程序 UI 的方式写一个运行在不同进程中的 Service。在 Android 平台中，一个进程通常不能访问其他进程中的内存区域。所以，它们需要把对象拆分成操作系统能理解的简单形式，以便伪装成对象跨越边界访问。编写这种伪装代码相当枯燥乏味，好在 Android 提供了 AIDL 工具来做这件事。

AIDL（Android Interface Language，Android 接口描述语言）是一个 IDL 语言，它可以生成一段代码，可以使在一个 Android 设备上运行的两个进程使用内部通信进程进行交互。如果需要在一个进程中（如在一个 Activity 中）访问另一个进程中（如一个 Service）某个对象的方法，可以使用 AIDL 生成这样的代码来伪装传递各种参数。

要使用 AIDL，Service 需要以 aidl 文件的方式提供服务接口，AIDL 工具将生成一个相应的 Java 接口，并且在生成的服务接口中包含一个功能调用的 stub 服务桩类。Service 的实现类需要继承这个 stub 服务桩类。Service 的 onBind()方法会返回实现类的对象，之后就可以使用它了。

3. 启动 Service 的两种方法

绑定 Activity 和 Service 的方式启动服务，通过 bindService 方法可以将 Activity 和 Service 绑定。bindService()方法的定义如下：

```
public boolean bindService(Intent service, ServiceConnection conn, int flags)
```

该方法的第一个参数表示与服务类相关联的 Intent 对象，第二个参数是一个 ServiceConnection 类型的变量，负责连接 Intent 对象指定的服务。通过 ServiceConnection 对象可以获得连接成功或失败的状态，并可以获得连接后的服务对象。第三个参数是一个标志位，一般设为 Context.BIND_AUTO_CREATE。

下面编写 MyService 类，在该类中增加了几个与绑定相关的事件方法。

```
public class MyService extends Service
{
    private MyBinder myBinder=new MyBinder();
    public IBinder onBind(Intent intent)
    {   //成功绑定后调用该方法
        Log.d("MyService", "onBind");
        return myBinder;
    }
     public void onRebind(Intent intent)
    {   //重新绑定时调用该方法
        Log.d("MyService", "onRebind");
        super.onRebind(intent);
    }
```

```
    public boolean onUnbind(Intent intent)
    {    //解除绑定时调用该方法
        Log.d("MyService", "onUnbind");
        return super.onUnbind(intent);
    }
    public void onCreate()
    {
        Log.d("MyService", "onCreate");
        super.onCreate();
    }
    public void onDestroy()
    {
        Log.d("MyService", "onDestroy");
        super.onDestroy();
    }
    public void onStart(Intent intent, int startId)
    {
        Log.d("MyService", "onStart");
        super.onStart(intent, startId);
    }
    public class MyBinder extends Binder
    {
        MyService getService()
        {
            return MyService.this;
        }
    }
}
```

现在定义一个 MyService 变量和一个 ServiceConnection 变量，代码如下：

```
private MyService myService;
private ServiceConnection serviceConnection=new ServiceConnection()
{    //连接服务失败后，该方法被调用
    public void onServiceDisconnected(ComponentName name)
    {
        myService=null;
        Toast.makeText(Main.this,"Service Failed.",Toast.LENGTH_LONG).show();
    }
    //成功连接服务后，该方法被调用。在该方法中可以获得MyService对象
    public void onServiceConnected(ComponentName name, IBinder service)
    {   //获得MyService对象
        myService=((MyService.MyBinder) service).getService();
        Toast.makeText(Main.this,"Service Connected.",Toast.LENGTH_LONG).show();
    }
};
```

最后使用 bindService()方法绑定 Activity 和 Service，代码如下：

```
bindService(serviceIntent, serviceConnection, Context.BIND_AUTO_CREATE);
```

如果想解除绑定，可以使用下面的代码：

```
unbindService(serviceConnection);
```

在 MyService 类中定义了一个 MyBinder 类，该类实际上是为了获得 MyService 对象的实例。在 ServiceConnection 接口的 onServiceConnected()方法的第二个参数是一个 IBinder 类型的变量，将该参数转换成 MyService.MyBinder 对象，并使用 MyBinder 类中的 getService 方法获得 MyService 对象。在获得 MyService 对象后，就可以在 Activity 中随意操作 MyService 了。

在 BroadcastReceiver 中启动 Service，先编写一个服务类，这个服务类没什么特别的，仍然使用前面编写的 MyService 类即可。在 AndroidManifest.xml 文件中配置 MyService 类的代码也相同。

下面完成关键一步，就是建立一个 BroadcastReceiver，代码如下：

```
public class StartupReceiver extends BroadcastReceiver
{
    public void onReceive(Context context, Intent intent)
    {   //启动一个 Service
        Intent serviceIntent=new Intent(context, MyService.class);
        context.startService(serviceIntent);
        Intent activityIntent=new Intent(context, MessageActivity.class);
        //要想在 Service 中启动 Activity，必须设置如下标志
        activityIntent.setFlags(Intent.FLAG_ACTIVITY_NEW_TASK);
        context.startActivity(activityIntent);
    }
}
```

在 StartupReceiver 类的 onReceive 方法中完成了两项工作：启动服务和显示一个 Activity 来提示服务启动成功。其中 MessageActivity 是一个普通的 Activity 类，只是该类在配置时使用了@android:style/Theme.Dialog 主题。

如果安装本例后，在重新启动模拟器后并未出现信息提示框，最大的可能是没有在 AndroidManifest.xml 文件中配置 BroadcastReceiver 和 Service，AndroidManifest.xml 文件的完整代码如下：

```
<?xml version="1.0" encoding="utf-8"?>
<manifest xmlns:android=http://schemas.android.com/apk/res/android
package="net.blogjava.mobile.startupservice" android:versionCode="1"
    android:versionName="1.0">
        <application  android:icon="@drawable/icon"   android:label="@string/
app_name">
            <activity android:name=".MessageActivity" android:theme="@android:
style/Theme.Dialog">
                <intent-filter>
                    <category android:name="android.intent.category.LAUNCHER" />
                </intent-filter>
            </activity>
            <receiver android:name="StartupReceiver">
                <intent-filter>
                    <action android:name="android. intent.action.BOOT_COMPLETED" />
                    <category android:name="android. intent.category.LAUNCHER" />
                </intent-filter>
            </receiver>
            <service android:enabled="true" android:name=".MyService" />
        </application>
```

```
    <uses-sdk android:minSdkVersion="3" />
    <uses-permission android:name="android. permission.RECEIVE_BOOT_COMPLETED" />
</manifest>
```

4. Service 典型应用案例

（1）Service 的启动和停止，具体的代码请见配套光盘中的 CH01_3。

该程序中包括两个按钮，一个按钮用于启动 Service，一个按钮用于关闭 Service。通过该实例，让用户了解 Service 启动和关闭的具体调用方法。该程序的界面布局如图 1-7 所示。

图 1-7　服务主界面

该程序的界面布局代码如下：CH01\CH01_3\res\layout\ activity_main.xml。

```
<?xml version="1.0" encoding="utf-8"?>
<LinearLayout xmlns:android="http://schemas.android.com/apk/res/android"
   android:orientation="horizontal"
   android:layout_width="fill_parent"
   android:layout_height="fill_parent"
   android:gravity="center_horizontal">
<Button
   android:id="@+id/start"
   android:layout_width="wrap_content"
   android:layout_height="wrap_content"
   android:text="@string/start" />
<Button
    android:id="@+id/stop"
    android:layout_width="wrap_content"
    android:layout_height="wrap_content"
    android:text="@string/stop" />
</LinearLayout>
```

程序的 Java 代码如下：CH01\CH01_3\src\com\zigcloud\service\StartServiceTest.java。

```
//从 Activity 基类派生子类 StartServiceDemo
public class StartServiceDemo extends Activity
{
   Button startBtn, stopBtn;                            //定义两个按钮对象
   private static final String TAG="StartServiceDemo";
                                                       //定义私有静态只读字符串变量
   public void onCreate(Bundle savedInstanceState) //重写子类 onCreate 方法
   {
      super.onCreate(savedInstanceState);             //调用基类 onCreate 方法
      setContentView(R.layout.activity_main);         //设置活动屏幕布局
      //获取程序界面中的 start、stop 两个按钮
      startBtn=(Button) findViewById(R.id.start);    //查找启动服务按钮对象
      stopBtn=(Button) findViewById(R.id.stop);    //查找停止服务按钮对象
      final Intent intent=new Intent();               //创建启动 Service 的 Intent
      intent.setAction("com.zigcloud.service.FIRST_SERVICE");
                                                      //为 Intent 设置 Action 属性
      startBtn.setOnClickListener(new OnClickListener()//设置按钮单击监听器
      {
         public void onClick(View arg0)               //处理按钮单击回调方法
```

```
            {
                Log.d(TAG, "启动服务!");
                startService(intent);                      //启动指定 Serivce
            }
        });
        stopBtn.setOnClickListener(new OnClickListener() //设置按钮单击监听器
        {
            public void onClick(View arg0)               //处理按钮单击回调方法
            {
                Log.d(TAG, "停止服务!");
                stopService(intent);                       //停止指定 Serivce
            }
        });
    }
}
```

程序清单 CH01\CH01_3\src\com\zigcloud\service\FirstService.java。

```
public class FirstService extends Service//从 Service 基类派生子类 FirstService
{
    private static final String TAG="第一个服务";
    public IBinder onBind(Intent arg0)    //必须实现的方法
    {
        return null;
    }
    public void onCreate()                  //Service 被创建时回调该方法
    {
        super.onCreate();
        Log.d(TAG, "服务已经被创建!");
    }
    // Service 被启动时回调该方法
    public int onStartCommand(Intent intent, int flags, int startId)
    {
        Log.d(TAG, "服务已经启动!");
        return START_STICKY;
    }
    public void onDestroy()                 //Service 被关闭之前回调
    {
        super.onDestroy();
        Log.d(TAG, "服务已经被销毁!");
    }
}
```

（2）Service 的绑定机制，具体代码请见配套光盘中的 CH01_4。

该程序通过在 Activity 中绑定本地 Service，来获取 Service 的运行状态。同时，通过该实例，让用户了解解除绑定的实现。该程序的界面布局如图 1-8 所示。通过 DDMS 可以观察到服务启动和停止过程。

图 1-8　系统服务案例界面

程序的界面布局代码如下：CH01\ CH01_4\res\layout\ activity_main.xml。

```
<?xml version="1.0" encoding="utf-8"?>
```

```
<LinearLayout xmlns:android="http://schemas.android.com/apk/res/android"
    android:orientation="horizontal"
    android:layout_width="fill_parent"
    android:layout_height="fill_parent"
    android:gravity="center_horizontal">
<Button
    android:id="@+id/bind"
    android:layout_width="wrap_content"
    android:layout_height="wrap_content"
    android:text="@string/bind"
    android:textSize="14dp"/>
<Button
    android:id="@+id/unbind"
    android:layout_width="wrap_content"
    android:layout_height="wrap_content"
    android:text="@string/unbind"
    android:textSize="14dp"/>
<Button
    android:id="@+id/getServiceStatus"
    android:layout_width="wrap_content"
    android:layout_height="wrap_content"
    android:text="@string/getServiceStatus"
    android:textSize="14dp"/>
</LinearLayout>
```

程序的 Java 代码如下：CH01\CH01_4\src\com\zigcloud\service\ BindService.java。

```
public class BindService extends Service
{
   private int count;
   private boolean quit;
   private MyBinder binder=new MyBinder();//定义 onBinder 方法所返回的对象
   public class MyBinder extends Binder  //通过继承 Binder 来实现 IBinder 类
   {
      public int getCount()
      {   //获取 Service 的运行状态：count
         return count;
      }
   }
   public IBinder onBind(Intent intent)
   {  //必须实现的方法，绑定该 Service 时回调该方法
      System.out.println("Service is Binded");
      return binder;              //返回 IBinder 对象
   }
   public void onCreate()
   {  //Service 被创建时回调该方法
      super.onCreate();
      System.out.println("Service is Created");
```

```
        new Thread()              //启动一条线程、动态地修改 count 状态值
        {
            public void run()
            {
                while (!quit)
                {
                    try
                    {
                        Thread.sleep(1000);
                    }
                    catch (InterruptedException e)
                    {
                    }
                    count++;
                }
            }
        }.start();
    }
    public boolean onUnbind(Intent intent)
    {   //Service 被断开连接时回调该方法
        System.out.println("Service is Unbinded");
        return true;
    }
    public void onDestroy()
    {   //Service 被关闭之前回调该方法
        super.onDestroy();
        this.quit=true;
        System.out.println("Service is Destroyed");
    }
}
```

程序清单 CH01\CH01_4\src\com\zigcloud\service\BindServiceDemo.java

```
public class BindServiceDemo extends Activity
{
    Button bind, unbind, getServiceStatus;
    BindService.MyBinder binder;         //保持所启动的 Service 的 IBinder 对象
    // 定义一个 ServiceConnection 对象
    private ServiceConnection conn=new ServiceConnection()
    {   //当该 Activity 与 Service 连接成功时回调该方法
        public void onServiceConnected(ComponentName name, IBinder service)
        {
            System.out.println("--Service Connected--");
            //获取 Service 的 onBind 方法所返回的 MyBinder 对象
            binder=(BindService.MyBinder) service;
        }
        //当该 Activity 与 Service 断开连接时回调该方法
        public void onServiceDisconnected(ComponentName name)
```

```
        {
            System.out.println("--Service Disconnected--");
        }
    };
    public void onCreate(Bundle savedInstanceState)
    {
        super.onCreate(savedInstanceState);
        setContentView(R.layout.activity_main);
        //获取程序界面中的 start、stop、getServiceStatus 按钮
        bind=(Button) findViewById(R.id.bind);
        unbind=(Button) findViewById(R.id.unbind);
        getServiceStatus=(Button) findViewById(R.id.getServiceStatus);
        final Intent intent=new Intent();     // 创建启动 Service 的 Intent
        intent.setAction("org.crazyit.service.BIND_SERVICE");//为 Intent 设
置 Action 属性
        bind.setOnClickListener(new OnClickListener()
        {
            public void onClick(View source)
            {  //绑定指定 Serivce
                bindService(intent, conn, Service.BIND_AUTO_CREATE);
            }
        });
        unbind.setOnClickListener(new OnClickListener()
        {
            public void onClick(View source)
            {  //解除绑定 Serivce
                unbindService(conn);
            }
        });
        getServiceStatus.setOnClickListener(new OnClickListener()
        {
            public void onClick(View source)
            {    //获取、并显示 Service 的 count 值
                Toast.makeText(BindServiceDemo.this, "Serivce 的 count 值为: "
+ binder.getCount(),Toast.LENGTH_SHORT).show();
            }
        });
    }
}
```

该程序的运行效果如图 1-9 所示。

图 1-9　系统服务案例运行效果

1.3 案例

1.3.1 Android 应用程序过渡页面

一些 Android 应用程序通常在启动时会显示过渡页面，显示过渡页面有几个好处：显示跟程序相关的一些介绍，在过渡页面呈现的过程中，初始化程序所需要的资源（检测网络状况，GPS 是否打开等），避免在初始化资源过程中出现卡顿等现象，改善用户体验。具体的代码请见配套光盘中的 CH01_5。该程序的界面布局如图 1-10 所示。

CH01_5_过渡页面案例

欢迎使用Android高级应用实战教程！

图 1-10　过渡页面案例布局主界面

程序的界布局代码如下：CH01\CH01_5\res\layout\ activity_main.xml。

```
<?xml version="1.0" encoding="utf-8"?>
<LinearLayout xmlns:android="http://schemas.android.com/apk/res/android"
    android:layout_width="fill_parent"
    android:layout_height="fill_parent"
    android:layout_gravity="center_vertical"
    android:background="@drawable/splash_bg"
    android:orientation="vertical"
    android:paddingBottom="7.0dip"
    android:paddingLeft="7.0dip"
    android:paddingRight="7.0dip"
    android:paddingTop="7.0dip" >
</LinearLayout>
```

程序的 Java 代码如下：CH01\CH01_5\src\com\example\demosplash\ SplashActivity.java。

```
public class SplashActivity extends Activity{    /*** 过渡页面 * */
  private Handler mHandler=new Handler();
  private final int SPLASH_TIME=500;
  public void onCreate(Bundle savedInstanceState) {
     super.onCreate(savedInstanceState);
     //setFullScreen();
     View view=View.inflate(this,R.layout.activity_splash, null);
                          //加载过渡页面的布局文件
     setContentView(view);
     Animation animation=AnimationUtils.loadAnimation(this, R.anim.myanim);
                          //定义过渡动画
     view.startAnimation(animation);
     animation.setAnimationListener(new AnimationListener() {
        public void onAnimationStart(Animation arg0) {}
        public void onAnimationRepeat(Animation arg0) {}
        public void onAnimationEnd(Animation arg0) {
           mHandler.postDelayed(new Runnable() {
              public void run() {
                goMain();
              }
           }, SPLASH_TIME);
        }
```

```
        });
    }
    protected void onResume() {
        super.onResume();
    }
    private void goMain() { /*** 跳转到主窗体* */
        startActivity(new Intent(this,MainActivity.class));
    };
    @SuppressWarnings("unused")
    private void setFullScreen(){    /*** 设置全屏* */
        getWindow().
setFlags(WindowManager.LayoutParams.FLAG_FULLSCREEN,WindowManager.LayoutParams.
FLAG_FULLSCREEN);
    }
    @SuppressWarnings("unused")
    private void quitFullScreen(){/*** 退出全屏* */
        final WindowManager.LayoutParams attrs = getWindow().getAttributes();
        attrs.flags &= (~WindowManager.LayoutParams.FLAG_FULLSCREEN);
        getWindow().setAttributes(attrs);
        getWindow().clearFlags(WindowManager.LayoutParams.FLAG_LAYOUT_NO_
LIMITS);
    }
}
```

程序清单 CH01\ CH01_5\src\ com\example\demosplash\MainActivity.java。

```
public class MainActivity extends Activity {/*** 主页面* */
    protected void onCreate(Bundle savedInstanceState) {
        super.onCreate(savedInstanceState);
        setContentView(R.layout.activity_main);
    }
    public boolean onCreateOptionsMenu (Menu menu) {
        getMenuInflater().inflate(R.menu.main, menu);
                                            //膨胀的菜单，将菜单项添加到行动条
        return true;
    }
}
```

该程序的运行效果如图 1-11 所示。

图 1-11　过渡页面案例运行主界面

1.3.2　服务器/客户端通信中的心跳包功能

在开发一些服务器/客户端程序时，需要定时与服务器通信或定时执行一些本地任务，在程序退出时，如果有消息更新，需要及时唤醒程序并对程序的界面进行更新，就需要使用到 Android 中的 Service 服务来实现此功能。具体的代码请见配套光盘中的 CH01_6。该程序的界面布局如图 1-12 所示。

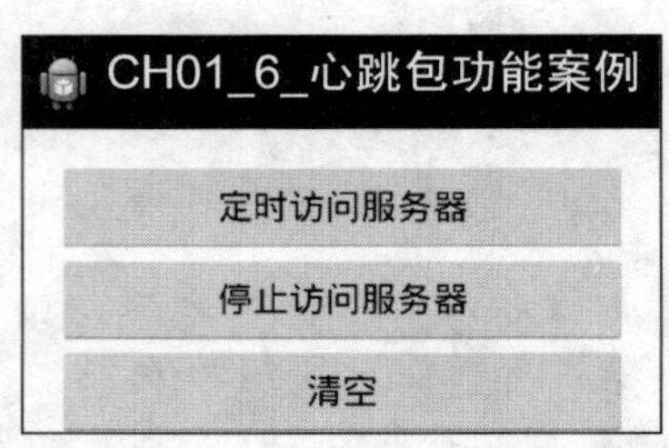

图 1-12　心跳包功能案例布局主界面

程序的界布局代码如下：CH01\CH01_6\res\layout\ activity_main.xml。

```
<RelativeLayout xmlns:android="http://schemas.android.com/apk/res/android"
    xmlns:tools="http://schemas.android.com/tools"
    android:id="@+id/RelativeLayout1"
    android:layout_width="match_parent"
    android:layout_height="match_parent"
    android:orientation="vertical"
    android:paddingBottom="@dimen/activity_vertical_margin"
    android:paddingLeft="@dimen/activity_horizontal_margin"
    android:paddingRight="@dimen/activity_horizontal_margin"
    android:paddingTop="@dimen/activity_vertical_margin"
    tools:context=".MainActivity" >
    <Button
        android:id="@+id/btn_connect"
        android:layout_width="match_parent"
        android:layout_height="wrap_content"
        android:layout_alignLeft="@+id/btn_disconnect"
        android:layout_alignParentTop="true"
        android:text="@string/connect" />
    <Button
        android:id="@+id/btn_disconnect"
        android:layout_width="match_parent"
        android:layout_height="wrap_content"
        android:layout_below="@+id/btn_connect"
        android:layout_centerHorizontal="true"
        android:text="@string/disconnect" />
    <Button
        android:id="@+id/btn_clear"
        android:layout_width="match_parent"
        android:layout_height="wrap_content"
        android:layout_alignLeft="@+id/btn_disconnect"
        android:layout_below="@+id/btn_connect"
        android:layout_marginTop="47dp"
        android:text="@string/clear" />
    <TextView
        android:id="@+id/tv_content"
        android:layout_width="wrap_content"
        android:layout_height="wrap_content"
        android:layout_alignParentBottom="true"
        android:layout_alignParentLeft="true"
```

```
        android:layout_alignParentRight="true"
        android:layout_below="@+id/btn_clear"
        android:text="" />
</RelativeLayout>
```

程序的 Java 代码如下：CH01\CH01_6\src\com\example\heart\MainActivity.java。

```
@SuppressLint("SimpleDateFormat")
public class MainActivity extends Activity {/***主窗体启动、停止心跳包服务**/
   private Button mConnect;
   private Button mDisConnect;
   private Button mClear;
   protected void onCreate(Bundle savedInstanceState) {
      super.onCreate(savedInstanceState);
      setContentView(R.layout.activity_main);
   }
   protected void onResume() {
      super.onResume();
      mConnect=(Button) findViewById(R.id.btn_connect);
      mConnect.setOnClickListener(mOnclickListener);
      mDisConnect=(Button) findViewById(R.id.btn_disconnect);
      mDisConnect.setOnClickListener(mOnclickListener);
      mClear=(Button) findViewById(R.id.btn_clear);
      mClear.setOnClickListener(mOnclickListener);
      addText(getResources().getString(R.string.updateTime)+":"
+getCurrentTime()+"\r\n");
   }
   private final OnClickListener mOnclickListener=new OnClickListener()
   {  /***按钮点击事件监听类**/
      public void onClick(View v) {
         switch (v.getId()) {
         case R.id.btn_connect:
            startService(new Intent(MainActivity.this, HeartService.class));
            addText(getResources().getString(R.string.connect));
            break;
         case R.id.btn_disconnect:
            stopService(new Intent(MainActivity.this, HeartService.class));
            addText(getResources().getString(R.string.disconnect));
            break;
         case R.id.btn_clear:
            clearText();
            break;
         default:
            break;
         }
      }
   };
   public boolean onCreateOptionsMenu(Menu menu) {
      getMenuInflater().inflate(R.menu.main, menu);
                                         //膨胀的菜单，将菜单项添加到行动条
```

```
        return true;
    }
    private void addText(final String str){ /*** 添加内容信息* */
        new Handler().post(new Runnable() {
            public void run() {
                TextView tv_content=(TextView)findViewById(R.id.tv_content);
                tv_content.setText(String.format("%s\r\n%s",tv_content.getText().
toString(), str));
            }
        });
    }
    private void clearText(){/*** 清空内容信息* */
        new Handler().post(new Runnable() {
            public void run() {
                TextView tv_content=(TextView)findViewById(R.id.tv_content);
                tv_content.setText("");
            }
        });
    }
    private String getCurrentTime(){/*** 获取系统当前时间* */
        SimpleDateFormat formatter=new SimpleDateFormat("yyyy/MM/dd HH:mm:ss");
        Date curDate=new Date(System.currentTimeMillis());
        String  str=formatter.format(curDate);
        return str;
    }
}
```

程序清单 CH01\CH01_6\src\ com\example\heart\receiver\HeartReceiver.java。

```
@SuppressLint("SimpleDateFormat")
public class HeartReceiver extends BroadcastReceiver {/*** 推送广播接收器*/
    private static final String TAG="HeartReceiver";
    public void onReceive(Context context, Intent intent) {
        String action=intent.getAction();
        Log.d(TAG, "action" + action);
        if (Const.ACTION_START_HEART.equals(action)) {
            Log.d(TAG, "Start heart");
        } else if (Const.ACTION_HEARTBEAT.equals(action)) {
            intent.setClass(context, MainActivity.class);
            intent.addFlags(Intent.FLAG_ACTIVITY_NEW_TASK);
            context.startActivity(intent);
            Log.d(TAG, "Heartbeat");
            //在此完成心跳需要完成的工作，比如请求远程服务器等
        } else if (Const.ACTION_STOP_HEART.equals(action)) {
            Log.d(TAG, "Stop heart");
        }
    }
}
```

程序清单 CH01\ CH01_6\src\ com\example\heart\service\ HeartService.java。

```
public class HeartService extends Service {/***心跳包服务**/
   private static final String TAG="HeartService";
   private static final long HEARTBEAT_INTERVAL=6*1000L;/***心跳间隔一分钟*/
   private AlarmManager mAlarmManager;
   private PendingIntent mPendingIntent;
   public IBinder onBind(Intent intent) {
      return null;
   }
   public void onCreate() {
      super.onCreate();
      mAlarmManager=(AlarmManager) getSystemService(ALARM_SERVICE);
      mPendingIntent=PendingIntent.getBroadcast(this, 0, new Intent(
Const.ACTION_HEARTBEAT), PendingIntent.FLAG_UPDATE_CURRENT);
   }
   public int onStartCommand(Intent intent, int flags, int startId) {
      Log.d(TAG, "onStartCommand");
      Intent startIntent=new Intent(Const.ACTION_START_HEART);
      //发送启动推送任务的广播
      sendBroadcast(startIntent);
      //启动心跳定时器
      long triggerAtTime=SystemClock.elapsedRealtime() + HEARTBEAT_INTERVAL;
      mAlarmManager.setInexactRepeating(AlarmManager.ELAPSED_REALTIME,
triggerAtTime, HEARTBEAT_INTERVAL, mPendingIntent);
      return super.onStartCommand(intent, flags, startId);
   }
   public void onDestroy() {
      Intent startIntent = new Intent(Const.ACTION_STOP_HEART);
      sendBroadcast(startIntent);
      mAlarmManager.cancel(mPendingIntent);    //取消心跳定时器
      super.onDestroy();
   }
}
```

程序清单 CH01\CH01_6\src\com\example\heart\utils\Const.java。

```
package com.example.heart.utils;
public interface Const {/***常量 */
   int PUSH_MSG=100;
   String ACTION_START_HEART="com.example.heart.intent.STARTPUSH";
   String   ACTION_HEARTBEAT="ccom.example.heart.
intent.HEARTBEAT";
   String   ACTION_STOP_HEART="com.example.heart.
intent.STOPPUSH";
}
```

该程序的运行效果如图 1-13 所示。

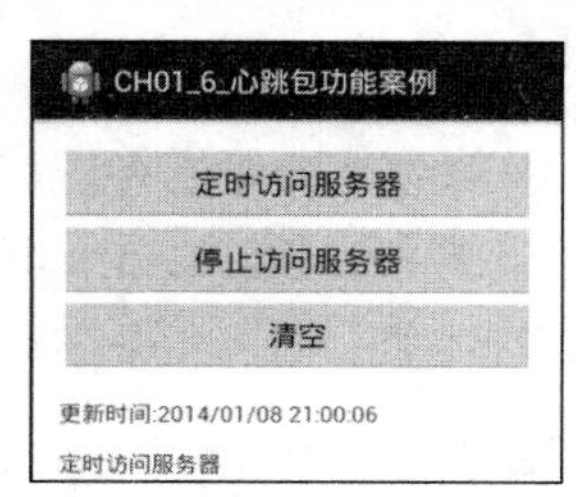

图 1-13　心跳包功能案例运行主界面

1.4 知识扩展

1.4.1 BroadcastReceiver（广播接收器）

BroadcastReceiver 是 Android 系统的四大组件之一，这种组件本质上是一种全局的监听器，用于监听系统全局的广播消息。BroadcastReceiver 启动的步骤：

（1）创建需要启动的 BroadcastReceiver 的 Intent。

（2）调用 Content 的 sendBroadcast() 或 sendOrderedBroadcast() 方法来启动指定的 BroadcastReceiver。

1.4.2 数据绑定 Bundle 的主要功能函数

Bundle 类是一个 final 类，其功能是应用于 Activity 之间相互传递值。其主要功能函数如下：

1. public void clear ()

功能：从这种捆绑的映射中移除所有元素。

2. public boolean containsKey (String key)

功能：如果给定的键是在这种捆绑的映射中，则返回 true。

参数：一个字符串键。

3. public Object get (String key)

功能：返回给定键的对象入口。

参数：一个字符串键。

4. public boolean getBoolean (String key)

功能：返回给定键的密钥，或值为相关联的默认值，如果没有所需的类型的映射关系，则给定键的存在值。

参数：一个字符串键。

5. public Bundle getBundle (String key)

功能：返回给定键的密钥，或者为 null，如果没有所需的类型映射为给定的密钥或一个空值是明确与该键关联的存在值。

参数：一个字符串键。

6. public String getString (String key)

功能：返回值与给定的键，或 NULL。

参数：一个字符串键。

7. public void putString (String key, String value)

功能：插入一个字符串值到这种捆绑的映射中，取代任何现有的值。

1.4.3 意图的主要功能函数

1. public Intent setClass (Context packageContext, Class<?> cls)

功能：调用 setComponent(ComponentName)返回一个类对象的名称。

2. public Intent putExtra (String name, String value)

功能：添加扩展数据到意图中。

1.5 本章小结

本章主要介绍了 Adnroid 系统中 Intent 和 Service 的功能和用法，并通过几个实例让用户掌握 Intent 和 Service 的使用。

1.6 强化练习

1. 填空题

（1）Android 四大基本组件分别是（　　）、（　　）、（　　）和（　　）。

（2）Service 启动到销毁的过程包括（　　）、（　　）和（　　）。

（3）Intent 元素包括（　　）、（　　）、（　　）、（　　）、（　　）和（　　）。

（4）Service 继承了（　　）、（　　）和（　　）三个方法。

（5）启动 Service 的两种方法（　　）和（　　）。

2. 问答题、操作题或编程题

（1）什么是意图（Intent）？

（2）什么是服务（Service）？

（3）如何实现 Android 平台组件之间的沟通？

（4）如何处理耗时的后台程序？

第2章 Android 数据永久存储应用

学习目标：

- 理解 SharedPreferences 的概念和作用；
- 理解 Android 的文件 IO；
- 理解 SQLite 数据库；
- 理解 ContentProvider 的功能与意义；
- 掌握 SharedPreferences 保存程序的参数、选项的方法；
- 掌握读、写 SD 卡上的文件的实现；
- 掌握 Android 的 API 操作 SQLite 数据库的方法；
- 掌握 ContentProvider 类的作用和常用方法的实现；
- 掌握网络存储数据的方法。

2.1 学习导入

Android 一共提供了以下五种不同的数据存储方式。

（1）使用 SharedPreferences 存储数据。

（2）文件存储数据。

（3）SQLite 数据库存储数据。

（4）使用 ContentProvider 存储数据。

（5）网络存储数据。

2.1.1 什么是 SharedPreferences（偏好数据存储）

SharedPreferences 是 Android 平台上一个轻量级的存储类，主要是针对系统配置信息的保存，比如给程序界面设置了音效，想在下一次启动时还能够保留上次设置的音效。由于 Android 系统的界面采用 Activity 栈的形式，当系统内存不足时会回收一些 Activity 界面，所以有些操作需要在不活动时保留下来，以便等再次激活时能够显示出来。又如，登录用户的用户名与密码，其采用了 Map 数据结构来存储数据，以键值的方式存储，可以简单的读取与写入。

2.1.2 什么是文件存储数据

顾名思义就是将要保存的数据以文件的形式保存。当需要调用所保存的数据时只要读取这些文件即可。需要注意，在 Android 中文件是 Linux 下的形式。

2.1.3 什么是 SQLite 数据库存储数据

Android 内嵌了功能比其他手持设备操作系统强大的关系型数据库 sqlite3，SQL 语句基本都可以使用，创建的数据可以用 adb shell 来操作。具体路径是/data/data/package_name/databases。SQLite 是一个开源的关系型数据库，与普通关系型数据库一样，也具有 ACID 的特性。

2.1.4 什么是 ContentProvider 存储数据

Android 提供了 ContentProvider，一个程序可以通过实现一个 ContentProvider 的抽象接口将自己的数据完全暴露出去，而且 ContentProviders 以类似数据库中表的方式将数据暴露，也就是说 ContentProvider 就像一个“数据库”。那么外界获取其提供的数据，也就应该与从数据库中获取数据的操作基本一样，只不过是采用 URI 表示外界需要访问的“数据库”。

2.1.5 什么是网络存储数据

Android 中的网络存储数据就是将数据通过网络保存在网络上。在实际使用中会使用 java.net.*和 android.net.*等类。

2.2 技术准备

2.2.1 SharedPreferences 存储数据

SharedPreferences是Android平台上一个轻量级的存储类，主要是保存一些常用的配置(如窗口状态)，一般在 Activity 中重载窗口状态时使用 SharedPreferences 完成，它提供了 Android 平台常规的 Long 长整形、Int 整形、String 字符串型的保存方式。

SharedPreferences 类似过去 Windows 系统上的 ini 配置文件，但是它分为多种权限，可以全局共享访问，最终是以 xml 方式来保存，从整体效率来看不是特别高，对于常规的轻量级而言比 SQLite 要好不少，如果存储量不大可以考虑自己定义文件格式。xml 处理时 Dalvik 会通过自带底层的本地 XML Parser 解析，比如 XMLpull 方式，这样对于内存资源占用比较好。

它的本质是基于 XML 文件存储 key-value 键值对数据，通常用来存储一些简单的配置信息。其存储位置在/data/data/<包名>/shared_prefs 目录下。

SharedPreferences 对象本身只能获取数据而不支持存储和修改，存储修改通过 Editor 对象实现。实现 SharedPreferences 存储的步骤如下：

（1）根据 Context 获取 SharedPreferences 对象。

（2）利用 edit()方法获取 Editor 对象。

（3）通过 Editor 对象存储 key-value 键值对数据。

（4）通过 commit()方法提交数据。

利用 SharedPreferences 实现系统配置信息存储的示例代码如下所示：

```
public class MainActivity extends Activity {
    public void onCreate(Bundle savedInstanceState) {
        super.onCreate(savedInstanceState);
```

```
        setContentView(R.layout.main);
        Context ctx=MainActivity.this;    //获取 SharedPreferences 对象
        SharedPreferences sp=ctx.getSharedPreferences("SP", MODE_PRIVATE);
        Editor editor=sp.edit();           //存入数据
        editor.putString("STRING_KEY", "string");
        editor.putInt("INT_KEY", 0);
        editor.putBoolean("BOOLEAN_KEY", true);
        editor.commit();
        Log.d("SP", sp.getString("STRING_KEY", "none"));
                                           //返回 STRING_KEY 的值
        Log.d("SP", sp.getString("NOT_EXIST", "none"));
                                   //如果 NOT_EXIST 不存在，则返回值为 none
    }
}
```

这段代码执行过后，即在/data/data/com.test/shared_prefs 目录下生成了一个 SP.xml 文件，一个应用可以创建多个这样的 xml 文件。

SharedPreferences 对象与 SQLite 数据库相比，免去了创建数据库，创建表，写 SQL 语句等诸多操作，相对而言更加方便，简洁。但是 SharedPreferences 也有其自身缺陷，比如其只能存储 Boolean、Int、Float、Long 和 String 五种简单的数据类型，比如其无法进行条件查询等。所以不论 SharedPreferences 的数据存储操作如何简单，它也只能是存储方式的一种补充，而无法完全替代如 SQLite 数据库等其他数据存储方式。

SharedPreferences 典型应用案例：记录应用程序的运行次数

该案例主要实现记录应用程序的调用次数，当用户第一次启动该应用程序时，系统创建 SharedPreferences 来记录程序调用次数，用户再次启动该应用程序时，系统先读取 SharedPreferences 中记录的调用次数，然后将调用次数增加 1。该程序的布局如图 2-1 所示，该案例的详细代码参见配套光盘中的 CH02_1。

图 2-1　数据存储案例布局主界面

程序的界面布局代码清单如下：CH02\CH02_1\res\layout\activity_main.xml。

```
<?xml version="1.0" encoding="utf-8"?>
<LinearLayout xmlns:android="http://schemas.android.com/apk/res/android"
   android:orientation="vertical"
   android:layout_width="fill_parent"
   android:layout_height="fill_parent">
<TextView
   android:layout_width="fill_parent"
   android:layout_height="wrap_content"
   android:text="@string/hello"/>
</LinearLayout>
```

该程序的 Java 代码：CH02\CH02_1\src\com\zigcloud\io\UseCount.java

```
public class UseCount extends Activity
{
   SharedPreferences preferences;
   public void onCreate(Bundle savedInstanceState)
   {
      super.onCreate(savedInstanceState);
```

```
        setContentView(R.layout.activity_main);
        preferences=getSharedPreferences("count", MODE_WORLD_READABLE);
        int count=preferences.getInt("count", 0);
                            //读取 SharedPreferences 里的 count 数据
        Toast.makeText(this, "程序以前被使用了" + count + "次。", Toast.LENGTH_
LONG).show();                                    //显示程序以前使用的次数
        Editor editor=preferences.edit();
        editor.putInt("count", ++count);         //存入数据
        editor.commit();                         //提交修改
    }
}
```

该程序的运行效果如图 2-2 所示。

图 2-2　数据存储案例运行主界面

2.2.2　文件存储数据

Activity 提供了 openFileOutput()方法，用于把数据输出到文件中，具体的实现过程与在 J2SE 环境中保存数据到文件中是一样的。文件可用来存放大量数据，如文本、图片、音频等。默认位置：/data/data/<包>/files/*.**。利用文件存储数据的示例代码如下：

```
public void save()
{
    try {
        FileOutputStream    outStream=this.openFileOutput("a.txt",Context.
MODE_WORLD_READABLE);
        outStream.write(text.getText().toString().getBytes());
        outStream.close();
        Toast.makeText(MyActivity.this,"Saved",Toast.LENGTH_LONG).show();
    } catch (FileNotFoundException e) {
        return;
    }
    catch (IOException e){
        return ;
    }
}
```

openFileOutput()方法的第一参数用于指定文件名称，不能包含路径分隔符“/”，如果文件不存在，Android 会自动创建它。创建的文件保存在/data/data/<package name>/files 目录，例如：/data/data/cn.itcast.action/files/itcast.txt，通过选择 Eclipse 菜单 Window→Show View→Other 命令，在弹出的窗口中展开 android 文件夹，选择下面的 File Explorer 视图，然后在 File Explorer 视图中展开/data/data/<package name>/files 目录就可以看到该文件。

openFileOutput()方法的第二参数用于指定操作模式，有四种模式，分别为：

```
Context.MODE_PRIVATE=0
Context.MODE_APPEND=32768
Context.MODE_WORLD_READABLE=1
Context.MODE_WORLD_WRITEABLE=2
```

Context.MODE_PRIVATE：为默认操作模式，代表该文件是私有数据，只能被应用本身访问，在该模式下，写入的内容会覆盖原文件的内容。如果想把新写入的内容追加到原文件中，

可以使用 Context.MODE_APPEND。

Context.MODE_APPEND：该模式会检查文件是否存在，存在就向文件追加内容，否则就创建新文件。

Context.MODE_WORLD_READABLE 和 Context.MODE_WORLD_WRITEABLE 用来控制其他应用是否有权限读/写该文件。

MODE_WORLD_READABLE：表示当前文件可以被其他应用读取。

MODE_WORLD_WRITEABLE：表示当前文件可以被其他应用写入。

如果希望文件被其他应用读和写，可以传入：openFileOutput("itcast.txt", Context.MODE_WORLD_READABLE+Context.MODE_WORLD_WRITEABLE); Android 有一套自己的安全模型，当应用程序（.apk）在安装时系统就会分配给它一个 userid，当该应用要去访问其他资源（如文件）时，就需要 userid 匹配。默认情况下，任何应用程序创建的文件 SharedPreferences 应该是私有的（位于/data/data/<package name>/files），其他程序无法访问。除非在创建时指定了 Context.MODE_WORLD_READABLE 或者 Context.MODE_WORLD_WRITEABLE，只有这样其他程序才能正确访问。读取文件示例：


```
public void load()
{
    try {
        FileInputStream inStream=this.openFileInput("a.txt");
        ByteArrayOutputStream stream=new ByteArrayOutputStream();
        byte[] buffer=new byte[1024];
        int length=-1;
        while((length=inStream.read(buffer))!=-1)    {
            stream.write(buffer,0,length);
        }
        stream.close();
        inStream.close();
        text.setText(stream.toString());
        Toast.makeText(MyActivity.this,"Loaded",Toast.LENGTH_LONG).show();
    } catch (FileNotFoundException e) {
        e.printStackTrace();
    }
    catch (IOException e){
        return ;
    }
}
```

私有文件只能被创建该文件的应用访问，如果希望文件能被其他应用读和写，可以在创建文件时，指定 Context.MODE_WORLD_READABLE 和 Context.MODE_WORLD_ WRITEABLE 权限。

Activity 还提供了 getCacheDir()和 getFilesDir()方法：getCacheDir()方法用于获取/data/data/<package name>/cache 目录；getFilesDir()方法用于获取/data/data/<package name>/files 目录。把文件存入 SDCard。

使用 Activity 的 openFileOutput()方法保存文件，文件是存放在手持设备空间上，一般手持设备的存储空间不是很大，存放些小文件还行，如果要存放像视频这样的大文件，是不可行的。对于像视频这样的大文件，可以把它存放在 SDCard。

创建 SDCard 可以在 Eclipse 创建模拟器的同时创建，也可以使用 DOS 命令进行创建，方法如下：

在 DOS 窗口中进入 Android SDK 安装路径的 tools 目录，输入以下命令创建一张容量为 2GB 的 SDCard，文件扩展名可以随便取，建议使用.img：mksdcard 2048M D:\AndroidTool\ sdcard.img，在程序中访问 SDCard 时，用户需要申请访问 SDCard 的权限。

在 AndroidManifest.xml 中加入访问 SDCard 的权限如下：

```
<!-- 在 SDCard 中创建与删除文件权限 -->
 <uses-permission android:name="android.permission.MOUNT_UNMOUNT_
FILESYSTEMS"/>
 <!-- 往 SDCard 写入数据权限 -->
<uses-permission android:name="android.permission.WRITE_EXTERNAL_
STORAGE"/>
```

要往 SDCard 中存放文件，程序必须先判断手持设备是否装有 SDCard，且可以进行读/写。

```
if(Environment.getExternalStorageState().equals(Environment.MEDIA_MOUNTED)){
    File sdCardDir=Environment.getExternalStorageDirectory();
                                               //获取 SDCard 目录
    File saveFile=new File(sdCardDir, "a.txt");
    FileOutputStream outStream=new FileOutputStream(saveFile);
    outStream.write("test".getBytes());
    outStream.close();
}
```

Environment.getExternalStorageState()方法用于获取 SDCard 的状态，如果手持设备装有 SDCard，并且可以进行读/写，那么该方法返回的状态为 Environment.MEDIA_MOUNTED。

Environment.getExternalStorageDirectory()方法用于获取 SDCard 的目录。当然要获取 SDCard 的目录，用户也可以编写如下代码：

```
File sdCardDir=new File("/sdcard");     //获取 SDCard 目录
File saveFile=new File(sdCardDir, "itcast.txt");
//上面两句代码可以合成一句:
File saveFile=new File("/sdcard/a.txt");
FileOutputStream outStream=new FileOutputStream(saveFile);
outStream.write("test".getBytes());
outStream.close();
```

文件存储数据典型应用案例：SD 卡文件浏览器

该案例利用 Java 的 File 类实现 SD 卡中文件的查看功能。当程序启动时，系统获取/mnt/sdcard 目录下的全部文件、文件夹，并使用 ListView 将它们显示出来。该程序的界面布局如图 2-3 所示，其详细代码参见配套光盘中的 CH02_2。

CH02_2_SD卡文件浏览器案例

图 2-3　数据 SD 卡文件浏览器案例布局主界面

程序的界面布局代码如下：CH02\CH02_2\res\layout\activity_main.xml。

```
<?xml version="1.0" encoding="utf-8"?>
<RelativeLayout xmlns:android="http://schemas.android.com/apk/res/android"
```

```
    android:layout_width="match_parent"
    android:layout_height="match_parent">
  <!-- 显示当前路径的文本框 -->
  <TextView
    android:id="@+id/path"
    android:layout_width="fill_parent"
    android:layout_height="wrap_content"
    android:layout_gravity="center_horizontal"
    android:layout_alignParentTop="true"/>
  <!-- 列出当前路径下所有文件的 ListView -->
  <ListView
    android:id="@+id/list"
    android:layout_width="wrap_content"
    android:layout_height="wrap_content"
    android:divider="#000"
    android:dividerHeight="1px"
    android:layout_below="@id/path"/>
  <!-- 返回上一级目录的按钮 -->
  <Button android:id="@+id/parent"
    android:layout_width="38dp"
    android:layout_height="34dp"
    android:background="@drawable/home"
    android:layout_centerHorizontal="true"
    android:layout_alignParentBottom="true"/>
  </RelativeLayout>
```

程序清单 CH02\CH02_2\res\layout\ line.xml。

```
<?xml version="1.0" encoding="UTF-8"?>
<LinearLayout xmlns:android="http://schemas.android.com/apk/res/android"
    android:orientation="horizontal"
    android:layout_width="fill_parent"
    android:layout_height="fill_parent">
<!-- 定义一个 ImageView，用于作为列表项的一部分-->
<ImageView android:id="@+id/icon"
    android:layout_width="40dp"
    android:layout_height="40dp"
    android:paddingLeft="10dp"/>
<!-- 定义一个 TextView，用于作为列表项的一部分 -->
<TextView android:id="@+id/file_name"
    android:layout_width="wrap_content"
    android:layout_height="wrap_content"
    android:textSize="16dp"
    android:gravity="center_vertical"
    android:paddingLeft="10dp"
    android:paddingTop="10dp"
    android:paddingBottom="10dp"/>
</LinearLayout>
```

程序的 Java 代码如下：CH02\CH02_2\src\com\zigcloud\io\ SDFileExplorer.java。

```
public class SDFileExplorer extends Activity
{
```

```
    ListView listView;
    TextView textView;
    File currentParent;         //记录当前的父文件夹
    File[] currentFiles;        //记录当前路径下的所有文件的文件数组
    public void onCreate(Bundle savedInstanceState)
    {
       super.onCreate(savedInstanceState);
       setContentView(R.layout.activity_main);
       listView=(ListView) findViewById(R.id.list);
                                //获取列出全部文件的 ListView
       textView=(TextView) findViewById(R.id.path);
       File root=new File("/mnt/sdcard/");       //获取系统的 SD 卡的目录
       if (root.exists())       //如果 SD 卡存在
       {
          currentParent=root;
          currentFiles=root.listFiles();
          inflateListView(currentFiles);
                                //使用当前目录下的全部文件、文件夹填充 ListView
       }
       listView.setOnItemClickListener(new OnItemClickListener()
       {  //ListView 的列表项的单击事件绑定监听器
          public void onItemClick(AdapterView<?> parent,View view,int position,
long id)
          {  //用户单击了文件，直接返回，不做任何处理
             if (currentFiles[position].isFile()) return;
             File[] tmp=currentFiles[position].listFiles();
                                //获取用户单击的文件夹下的所有文件
             if(tmp==null || tmp.length==0)
             {
                Toast.makeText(SDFileExplorer.this, "当前路径不可访问或该路径
下没有文件", Toast.LENGTH_SHORT).show();
             }
             else
             {  //获取用户单击的列表项对应的文件夹，设为当前的父文件夹
                currentParent=currentFiles[position];
                currentFiles=tmp;       //保存当前的父文件夹内的全部文件和文件夹
                inflateListView(currentFiles);     //再次更新 ListView
             }
          }
       });
       Button parent=(Button) findViewById(R.id.parent);
                                      //获取上一级目录的按钮
       parent.setOnClickListener(new OnClickListener()
       {
          public void onClick(View source)
          {
             try
```

```
            {
               if (!currentParent.getCanonicalPath().equals("/mnt/sdcard"))
               {   //获取上一级目录
                  currentParent=currentParent.getParentFile();
                  currentFiles=currentParent.listFiles();
                                          //列出当前目录下所有文件
                  inflateListView(currentFiles);     //再次更新 ListView
               }
            }
            catch(IOException e)
            {
               e.printStackTrace();
            }
         }
      });
   }
   private void inflateListView(File[] files)
   {   //创建一个 List 集合，List 集合的元素是 Map
      List<Map<String,Object>> listItems=new ArrayList<Map<String, Object>>();
      for(int i=0; i<files.length; i++)
      {
         Map<String, Object> listItem=new HashMap<String, Object>();
         if (files[i].isDirectory())
         {   //如果当前 File 是文件夹，使用 folder 图标；否则使用 file 图标
            listItem.put("icon", R.drawable.folder);
         }
         else
         {
            listItem.put("icon", R.drawable.file);
         }
         listItem.put("fileName", files[i].getName());
         listItems.add(listItem);      //添加 List 项
      }
      //创建一个 SimpleAdapter
     SimpleAdapter simpleAdapter=new SimpleAdapter(this,listItems, R.layout.
line,new String[]{"icon","fileName"},new int[]{R.id.icon, R.id.file_name });
      listView.setAdapter(simpleAdapter); //为 ListView 设置 Adapter
      try
      {
         textView.setText("当前路径为: " + currentParent.getCanonicalPath());
      }
      catch (IOException e)
      {
         e.printStackTrace();
      }
   }
}
```

该程序的运行效果如图 2-4 所示。

图 2-4　数据 SD 卡浏览器案例运行主界面

2.2.3　SQLite 数据库存储数据

SQLite 是轻量级嵌入式数据库引擎，它支持 SQL 语句，并且只利用很少的内存就有很好的性能。此外它还是开源的，任何人都可以使用它。许多开源项目（Mozilla、PHP、Python）都使用了 SQLite。SQLite 由以下几个组件组成：SQL 编译器、内核、后端以及附件。SQLite 通过利用虚拟机和虚拟数据库引擎（Virtual Datebase Engine，VDBE），使调试、修改和扩展 SQLite 的内核变得更加方便。其主要特点如下：

（1）面向资源有限的设备。

（2）没有服务器进程。

（3）所有数据存放在同一文件中跨平台。

（4）可自由复制。

SQLite 内部结构如图 2-5 所示。

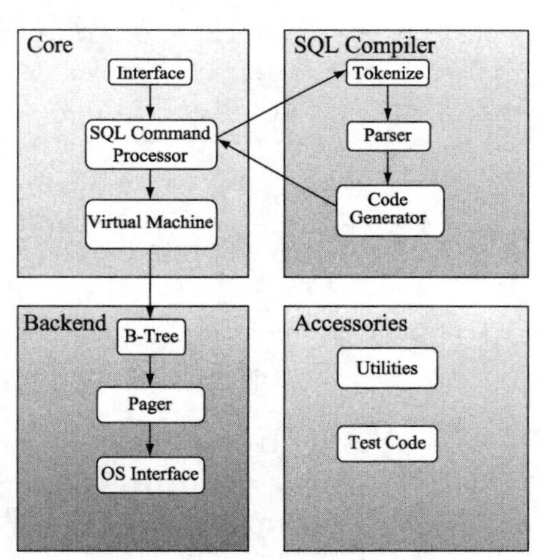

图 2-5　SQLite 内部结构

SQLite 基本上符合 SQL-92 标准，和其他的主要 SQL 数据库没什么区别。它的优点就是高效，Android 运行时环境包含了完整的 SQLite。

SQLite 和其他数据库最大的不同就是对数据类型的支持，创建一个表时，可以在 CREATE TABLE 语句中指定某列的数据类型，但是也可以把任何数据类型放入任何列中。当某个值插入数据库时，SQLite 将检查它的类型。如果该类型与关联的列不匹配，则 SQLite 会尝试将该值转换成该列的类型。如果不能转换，则该值将作为其本身具有的类型存储。比如可以把一个字符串（String）放入 INTEGER 列。SQLite 称这为“弱类型”（manifest typing）。此外，SQLite 不支持一些标准的 SQL 功能，特别是外键约束（FOREIGN KEY constrains），嵌套 transcaction 和 RIGHT OUTER JOIN 和 FULL OUTER JOIN，还有一些 ALTER TABLE 功能。除了上述功能外，SQLite 是一个完整的 SQL 系统，拥有完整的触发器、事务等。

Android 集成了 SQLite 数据库，所以每个 Android 应用程序都可以使用 SQLite 数据库。

对于熟悉 SQL 的开发人员来说，在 Android 开发中使用 SQLite 相当简单。但是，由于 JDBC 会消耗太多的系统资源，所以 JDBC 对于手持设备这种内存受限设备来说并不合适。因此，Android 提供了一些新的 API 来使用 SQLite 数据库，Android 开发中，程序员需要学会使用这些 API。

数据库存储在 data/< 项目文件夹 >/databases/ 下。Android 开发中使用 SQLite 数据库 Activites 可以通过 Content Provider 或者 Service 访问一个数据库。

下面详细讲解如何创建数据库，添加数据和查询数据库。

创建数据库，Android 不能自动提供数据库。在 Android 应用程序中使用 SQLite，必须自己创建数据库，然后创建表、索引，填充数据。

Android 提供 SQLiteOpenHelper 帮助用户创建数据库，用户只需继承 SQLiteOpenHelper 类即可轻松创建数据库。SQLiteOpenHelper 类根据开发应用程序的需要，封装了创建和更新数据库使用的逻辑。

SQLiteOpenHelper 的子类，至少需要实现以下三个方法：

（1）构造函数，调用父类 SQLiteOpenHelper 的构造函数。这个方法需要四个参数：上下文环境（如一个 Activity）、数据库名字、一个可选的游标工厂（通常是 Null）、一个代表用户正在使用的数据库模型版本的整数。

（2）onCreate()方法，它需要一个 SQLiteDatabase 对象作为参数，根据需要对这个对象填充表和初始化数据。

（3）onUpgrage() 方法，它需要三个参数，一个 SQLiteDatabase 对象、一个旧的版本号和一个新的版本号，这样用户就可以清楚如何把一个数据库从旧的模型转变到新的模型。

下面的示例代码展示了如何继承 SQLiteOpenHelper 创建数据库：

```
public class DatabaseHelper extends SQLiteOpenHelper {
    DatabaseHelper(Context  context,  String  name,  CursorFactory
cursorFactory, int version){
        super(context, name, cursorFactory, version);
    }
    public void onCreate(SQLiteDatabase db) {
      //TODO 创建数据库后，对数据库的操作
    }
    public void onUpgrade(SQLiteDatabase db,int oldVersion,int newVersion) {
      //TODO 更改数据库版本的操作
    }
    public void onOpen(SQLiteDatabase db) {
        super.onOpen(db);
        //TODO 每次成功打开数据库后首先被执行
    }
}
```

接下来讨论具体如何创建表、插入数据、删除表等。调用 getReadableDatabase()方法或 getWriteableDatabase()方法，用户可以得到 SQLiteDatabase 实例，具体调用哪个方法，取决于用户是否需要改变数据库的内容：

```
db=(new DatabaseHelper(getContext())).getWritableDatabase();
return (db==null) ? false : true;
```

上面这段代码会返回一个 SQLiteDatabase 类的实例，使用这个对象，就可以查询或者修改数据库。当完成了对数据库的操作（如 Activity 已经关闭），需要调用 SQLiteDatabase 的 Close() 方法释放数据库连接。创建表和索引，需要调用 SQLiteDatabase 的 execSQL() 方法来执行 DDL 语句。如果没有异常，这个方法没有返回值。例如，执行如下代码：

```
db.execSQL("CREATE TABLE mytable (_id INTEGER PRIMARY KEY AUTOINCREMENT
, title TEXT, value REAL);");
```

这条语句会创建一个名为 mytable 的表，表有一个列名为 _id，并且是主键，这列的值是会自动增长的整数（例如，当插入一行时，SQLite 会给这列自动赋值）。另外，还有两列：title（字符）和 value（浮点数）。SQLite 会自动为主键列创建索引。通常情况下，第一次创建数据库时创建了表和索引。

如果不需要改变表的 schema，不需要删除表和索引。删除表和索引，需要使用 execSQL() 方法调用 DROP INDEX 和 DROP TABLE 语句。给表添加数据，上面的代码，已经创建了数据库和表，现在需要给表添加数据。有两种方法可以给表添加数据。

像上面创建表一样，可以使用 execSQL()方法执行 INSERT、UPDATE、DELETE 等语句来更新表的数据。execSQL() 方法适用于所有不返回结果的 SQL 语句。例如：

```
db.execSQL("INSERT INTO widgets (name, inventory)"+ "VALUES ('Sprocket', 5)");
```

另一种方法是使用 SQLiteDatabase 对象的 insert()、update()、delete() 方法。这些方法把 SQL 语句的一部分作为参数。示例如下：

```
ContentValues cv=new ContentValues();
cv.put(Constants.TITLE, "example title");
cv.put(Constants.VALUE, SensorManager.GRAVITY_DEATH_STAR_I);
db.insert("mytable", getNullColumnHack(), cv);
```

update()方法有四个参数，分别是表名，表示列名和值的 ContentValues 对象，可选的 WHERE 条件和可选的填充 WHERE 语句的字符串，这些字符串会替换 WHERE 条件中的“?”标记。

update()根据条件更新指定列的值，所以用 execSQL() 方法可以达到同样的目的。WHERE 条件和其参数和用过的其他 SQL APIs 类似。例如：

```
String[] parms=new String[] {"this is a string"};
db.update("widgets", replacements, "name=?", parms);
```

delete() 方法的使用和 update()方法类似，使用表名，可选的 WHERE 条件和相应的填充 WHERE 条件的字符串。 查询数据库，类似 INSERT、UPDATE、DELETE，有两种方法使用 SELECT 从 SQLite 数据库检索数据。

使用 rawQuery()方法直接调用 SELECT 语句；使用 query()方法构建一个查询。

Raw Queries 正如 API 名字，rawQuery()是最简单的解决方法。通过这个方法可以调用 SQL SELECT 语句。例如：

```
Cursor c=db.rawQuery( "SELECT name FROM sqlite_master WHERE type='table'
AND name='mytable'", null);
```

在上面的例子中，查询 SQLite 系统表（sqlite_master）检查 table 表是否存在。返回值是一个 cursor 对象，这个对象的方法可以迭代查询结果。如果查询是动态的，使用这个方法就会非常复杂。例如，当需要查询的列在程序编译的时候不能确定，这时候使用 query()方法会方便很多。

Regular Queries query()方法用 SELECT 语句段构建查询。SELECT 语句内容作为 query() 方法的参数，例如：要查询的表名，要获取的字段名，WHERE 条件，包含可选的位置参数，去替代 WHERE 条件中位置参数的值，GROUP BY 条件，HAVING 条件。 除了表名，其他

参数可以是 null。所以，以前的代码段可以写成：

```
String[] columns={"ID", "inventory"};
String[] parms={"snicklefritz"};
Cursor result=db.query("widgets", columns, "name=?",parms, null, null, null);
```

使用游标，不管如何执行查询，都会返回一个 Cursor，这是 Android 的 SQLite 数据库游标。使用游标，可以：

（1）通过使用 getCount()方法得到结果集中有多少记录。

（2）通过 moveToFirst()、moveToNext()和 isAfterLast()方法遍历所有记录。

（3）通过 getColumnNames()方法得到字段名。

（4）通过 getColumnIndex()方法转换成字段号。

（5）通过 getString()、getInt()等方法得到给定字段当前记录的值。

（6）通过 requery()方法重新执行查询得到游标。

（7）通过 close()方法释放游标资源。

下面代码遍历 mytable 表：

```
Cursor result=db.rawQuery("SELECT ID, name, inventory FROM mytable");
result.moveToFirst();
while (!result.isAfterLast()) {
    int id=result.getInt(0);
    String name=result.getString(1);
    int inventory=result.getInt(2);
    result.moveToNext();
}
result.close();
```

在 Android 中使用 SQLite 数据库管理工具。在其他数据库上进行开发时，一般都使用工具来检查和处理数据库的内容，而不是仅仅使用数据库的 API。

使用 Android 模拟器，有两种可供选择的方法来管理数据库。

首先，模拟器绑定了 sqlite3 控制台程序，可以使用 adb shell 命令来调用它。只要进入了模拟器的 shell，在数据库的路径执行 sqlite3 命令就可以了。数据库文件一般存放在：/data/data/your.app.package/databases/your-db-name。如果喜欢使用更友好的工具，可以把数据库复制到开发机上，使用 SQLite-aware 客户端来操作它。这样的话，在一个数据库的副本上操作，如果想要修改能反映到设备上，需要把数据库备份回去。把数据库从设备上复制出来，可以使用 adb pull 命令（或者在 IDE 上进行相应操作）。存储一个修改过的数据库到设备上，使用 adb push 命令。一个最方便的 SQLite 客户端是 FireFox SQLite Manager 扩展，它可以跨所有平台使用。如果想要开发 Android 应用程序，一定需要在 Android 上存储数据，使用 SQLite 数据库是一种非常好的选择。

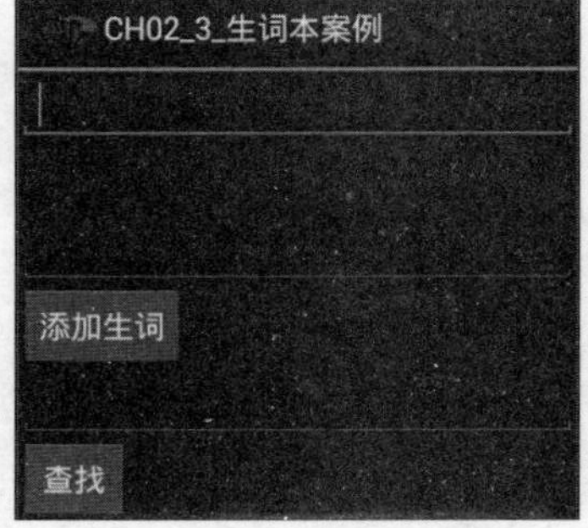

图 2-6　生词本案例布局主界面

SQLite 数据库存储数据典型应用案例：生词本

该案例实现的功能：用户可以将自己不熟悉的单词加到系统数据库中，当用户需要查询某个单词或解译时，只要在程序中输入相应的关键词，程序中相应的条目就可以显示出来。该程序的界面布局如图 2-6 所示，其详细代码请参见配套光盘中的 CH02_3。

程序的界面布局代码如下：CH02\CH02_3\res\layout\ activity_main.xml。

```
<?xml version="1.0" encoding="utf-8"?>
<LinearLayout xmlns:android="http://schemas.android.com/apk/res/android"
  android:orientation="vertical"
  android:layout_width="fill_parent"
  android:layout_height="fill_parent">
<EditText
  android:id="@+id/word"
  android:layout_width="fill_parent"
  android:layout_height="wrap_content" />
<EditText
  android:id="@+id/detail"
  android:layout_width="fill_parent"
  android:layout_height="wrap_content"
  android:lines="3"/>
<Button
  android:id="@+id/insert"
  android:layout_width="wrap_content"
  android:laycut_height="wrap_content"
  android:text="@string/insert"/>
<EditText
  android:id="@+id/key"
  android:layout_width="fill_parent"
  android:layout_height="wrap_content" />
<Button
  android:id="@+id/search"
  android:layout_width="wrap_content"
  android:layout_height="wrap_content"
  android:text="@string/search"/>
<ListView
  android:id="@+id/show"
  android:layout_width="fill_parent"
  android:layout_height="fill_parent" />
</LinearLayout>
```

程序清单 CH02\CH02_3\res\layout\ line.xml。

```
<?xml version="1.0" encoding="utf-8"?>
<LinearLayout xmlns:android="http://schemas.android.com/apk/res/android"
  android:orientation="vertical"
  android:layout_width="fill_parent"
  android:layout_height="fill_parent">
<EditText
  android:id="@+id/word"
  android:layout_width="wrap_content"
  android:layout_height="wrap_content"
  android:width="120px"
  android:editable="false"/>
<TextView
  android:layout_width="fill_parent"
```

```
    android:layout_height="wrap_content"
    android:text="@string/detail"/>
  <EditText
    android:id="@+id/detail"
    android:layout_width="fill_parent"
    android:layout_height="wrap_content"
    android:editable="false"
    android:lines="3"/>
</LinearLayout>
```

程序清单：CH02\CH02_3\res\layout\ popup.xml。

```
<?xml version="1.0" encoding="utf-8"?>
<LinearLayout xmlns:android="http://schemas.android.com/apk/res/android"
    android:orientation="vertical"
    android:layout_width="fill_parent"
    android:layout_height="fill_parent"
    android:gravity="center"
    android:divider="?android:attr/dividerVertical"
    android:showDividers="beginning"
    android:dividerPadding="2dp">
<ListView
    android:id="@+id/show"
    android:layout_width="fill_parent"
    android:layout_height="fill_parent" />
</LinearLayout>
```


程序的 Java 代码如下：CH02\CH02_3\src\com\zigcloud\db\Dict.java。

```
public class Dict extends Activity
{
    MyDatabaseHelper dbHelper;
    Button insert=null;
    Button search=null;
    public void onCreate(Bundle savedInstanceState)
    {
        super.onCreate(savedInstanceState);
        setContentView(R.layout.activity_main);
        //创建 MyDatabaseHelper 对象，指定数据库版本为 1，此处使用相对路径即可,
        //数据库文件会自动保存在程序的数据文件夹的 databases 目录下
        dbHelper=new MyDatabaseHelper(this, "myDict.db3", 1);
        insert=(Button) findViewById(R.id.insert);
        search=(Button) findViewById(R.id.search);
        insert.setOnClickListener(new OnClickListener()
        {
            public void onClick(View source)
            {   //获取用户输入
                String word=((EditText)findViewById(R.id.word)).getText().toString();
                String detail=((EditText) findViewById(R.id.detail)).getText().toString();
                insertData(dbHelper.getReadableDatabase(), word, detail);
                                                //插入生词记录
                Toast.makeText(Dict.this, "添加生词成功！", 8000).show();
                                                //显示提示信息
```

```
            }
        });
        search.setOnClickListener(new OnClickListener()
        {
            public void onClick(View source)
            {   //获取用户输入
                String key=((EditText)findViewById(R.id.key)).getText(). toString();
                //执行查询
                Cursor cursor=dbHelper.getReadableDatabase().rawQuery(
"select * from dict where word like ? or detail like ?",new String[] { "%" +
key + "%", "%" + key + "%" });
                Bundle data=new Bundle();        //创建一个 Bundle 对象
                data.putSerializable("data", converCursorToList(cursor));
                Intent intent=new Intent(Dict.this, ResultActivity.class);
                                                 //创建一个 Intent
                intent.putExtras(data);
                startActivity(intent);           //启动 Activity
            }
        });
    }
    protected ArrayList<Map<String,String>>converCursorToList(Cursor cursor)
    {
        ArrayList<Map<String,String>>result=new ArrayList<Map<String, String>>();
        while(cursor.moveToNext())               //遍历 Cursor 结果集
        {   //将结果集中的数据存入 ArrayList 中
            Map<String, String> map=new HashMap<String, String>();
            //取出查询记录中第 2 列、第 3 列的值
            map.put("word", cursor.getString(1));
            map.put("detail", cursor.getString(2));
            result.add(map);
        }
        return result;
    }
    private void insertData(SQLiteDatabase db, String word, String detail)
    {   //执行插入语句
        db.execSQL("insert into dict values(null , ? , ?)", new String[] {word, detail });
    }
    public void onDestroy()
    {
        super.onDestroy();
        //退出程序时关闭 MyDatabaseHelper 里的 SQLiteDatabase
        if(dbHelper != null)
        {
            dbHelper.close();
        }
    }
}
```

程序清单：CH02\ CH02_3\src\com\zigcloud\db\ MyDatabaseHelper.java。

```
public class MyDatabaseHelper extends SQLiteOpenHelper
{
```

```
    final String CREATE_TABLE_SQL =
"create table dict(_id integer primary " + "key autoincrement , word , detail)";
    public MyDatabaseHelper(Context context, String name, int version)
    {
        super(context, name, null, version);
    }
    public void onCreate(SQLiteDatabase db)
    {  // 第一次使用数据库时自动建表
        db.execSQL(CREATE_TABLE_SQL);
    }
    public void onUpgrade(SQLiteDatabase db, int oldVersion, int newVersion)
    {
        System.out.println("---调用 onUpdate---+oldVersion+"--->"+newVersion);
    }
}
```

程序清单：CH02\ CH02_3\src\com\zigcloud\db\ ResultActivity.java。

```
public class ResultActivity extends Activity
{
    public void onCreate(Bundle savedInstanceState)
    {
        super.onCreate(savedInstanceState);
        setContentView(R.layout.popup);
        ListView listView=(ListView) findViewById(R.id.show);
        Intent intent=getIntent();
        Bundle data=intent.getExtras();    //获取该 intent 所携带的数据
        //从 Bundle 数据包中取出数据
        @SuppressWarnings("unchecked")
        List<Map<String,  String>>  list=(List<Map<String,  String>>)data.
getSerializable("data");
        //将 List 封装成 SimpleAdapter
        SimpleAdapter adapter=new SimpleAdapter(ResultActivity.this, list,
R.layout.line,
        new String[] { "word", "detail" }, new int[] {R.id.word, R.id.detail });
        listView.setAdapter(adapter);          //填充 ListView
    }
}
```

该程序的运行效果图如图 2-7 所示。

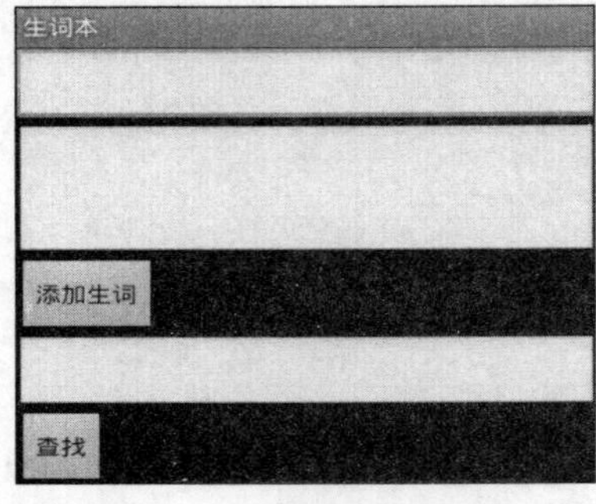

图 2-7　生词本案例运行主界面

2.2.4 ContentProvider 存储数据

Android 系统和其他的操作系统不太一样，需要记住的是，数据在 Android 当中是私有的，当然这些数据包括文件数据和数据库数据以及一些其他类型的数据。那这个时候有读者就会提出问题，难道两个程序之间就没有办法对数据进行交换？Android 这么优秀的系统不会让这种情况发生的。解决这个问题主要靠 ContentProvider。一个 ContentProvider 类实现了一组标准的方法接口，从而能够让其他的应用保存或读取此 ContentProvider 的各种数据类型。也就是说，一个程序可以通过实现一个 ContentProvider 的抽象接口将自己的数据暴露出去。外界根本看不到，也不用看到这个应用暴露的数据在应用当中是如何存储的，或者是用数据库存储还是用文件存储，还是通过网上获得，这些都不重要，重要的是外界可以通过这套标准及统一的接口和程序里的数据打交道，可以读取程序的数据，也可以删除程序的数据，当然，中间也会涉及一些权限的问题。

ContentProvider 提供了一种多应用间数据共享的方式，比如：联系人信息可以被多个应用程序访问。

ContentProvider 是个实现了一组用于提供其他应用程序存取数据的标准方法的类。应用程序可以在 ContentProvider 中执行如下操作：查询数据、修改数据、添加数据、删除数据。

标准的 ContentProvider：Android 提供了一些已经在系统中实现的标准 ContentProvider，比如联系人信息、图片库等，可以用这些 ContentProvider 访问设备上存储的联系人信息、图片等。

1. 查询记录

在 ContentProvider 中使用的查询字符串有别于标准的 SQL 查询。很多诸如 select、add、delete、modify 等操作都使用一种特殊的 URI（通用资源标识符，Uniform Resource Identifier）来进行，这种 URI 由三部分组成，“content://”代表数据的路径和一个可选的标识数据的 ID。以下是一些示例 URI：

```
content://media/internal/images    这个 URI 将返回设备上存储的所有图片
content://contacts/people/         这个 URI 将返回设备上的所有联系人信息
content://contacts/people/45       这个 URI 返回联系人信息中 ID 为 45 的联系人记录
```

尽管这种查询字符串格式很常见，但是它看起来还是有点令人迷惑。为此，Android 提供了一系列的帮助类（在 android.provider 包下），里面包含了很多以类变量形式给出的查询字符串，这种方式更容易让人理解，参见下例：

```
MediaStore.Images.Media.INTERNAL_CONTENT_URI
Contacts.People.CONTENT_URI
```

因此，如上面 content://contacts/people/45 这个 URI 就可以写成如下形式：

```
Uri person=ContentUris.withAppendedId(People.CONTENT_URI, 45);
```

然后执行数据查询：

```
Cursor cur=managedQuery(person, null, null, null);
```

这个查询返回一个包含所有数据字段的游标，可通过迭代这个游标来获取所有的数据：

```
public class ContentProviderDemo extends Activity {
    public void onCreate(Bundle savedInstanceState) {
        super.onCreate(savedInstanceState);
        setContentView(R.layout.main);
```

```
        displayRecords();
    }
    private void displayRecords() {
        //该数组中包含了所有要返回的字段
        String columns[]=new String[] { People.NAME, People.NUMBER };
        Uri mContacts=People.CONTENT_URI;
        Cursor cur=managedQuery(
            mContacts,
            columns,  // 要返回的数据字段
            null,          // WHERE 子句
            null,         // WHERE 子句的参数
            null          // Order-by 子句
        );
        if (cur.moveToFirst()) {
            String name=null;
            String phoneNo=null;
            do {  //获取字段的值
               name=cur.getString(cur.getColumnIndex(People.NAME));
               phoneNo=cur.getString(cur.getColumnIndex(People.NUMBER));
               Toast.makeText(this,name+" "+phoneNo, Toast.LENGTH_LONG).show();
          } while (cur.moveToNext());
        }
    }
}
```

上例示范了一个如何依次读取联系人信息表中的指定数据列 name 和 number。

2. 修改记录

```
private void updateRecord(int recNo, String name) {
    Uri uri=ContentUris.withAppendedId(People.CONTENT_URI, recNo);
    ContentValues values=new ContentValues();
    values.put(People.NAME, name);
    getContentResolver().update(uri, values, null, null);
}
```

现在可以调用上面的方法来更新指定记录：

```
updateRecord(10, "XYZ"); //更改第 10 条记录的 name 字段值为 XYZ。
```

3. 添加记录

要增加记录，可以调用 ContentResolver.insert()方法，该方法接受一个要增加的记录的目标 URI，以及一个包含了新记录值的 Map 对象，调用后的返回值是新记录的 URI，包含记录号。上面的例子中都是基于联系人信息簿这个标准的 ContentProvider，现在继续来创建一个 insertRecord()方法对联系人信息簿添加数据：

```
private void insertRecords(String name, String phoneNo) {
ContentValues values=new ContentValues();
    values.put(People.NAME, name);
    Uri uri=getContentResolver().insert(People.CONTENT_URI, values);
    Log.d("ANDROID", uri.toString());
    Uri numberUri=Uri.withAppendedPath(uri,People.Phones.CONTENT_DIRECTORY);
    values.clear();
    values.put(Contacts.Phones.TYPE, People.Phones.TYPE_MOBILE);
    values.put(People.NUMBER, phoneNo);
```

```
    getContentResolver().insert(numberUri, values);
}
```

这样就可以调用 insertRecords(name,phoneNo)的方式来向联系人信息簿中添加联系人姓名和电话号码。

4. **删除记录**

ContentProvider 中的 getContextResolver.delete()方法可以用来删除记录。

下面的记录用来删除设备上所有的联系人信息：

```
private void deleteRecords() {
    Uri uri=People.CONTENT_URI;
    getContentResolver().delete(uri, null, null);
}
```

也可以指定 WHERE 条件语句来删除特定的记录：

```
getContentResolver().delete(uri, "NAME="+"'XYZ XYZ'", null);
```

这将会删除 name 为 XYZ XYZ 的记录。

创建 ContentProvider:

至此已经知道如何使用 ContentProvider 了，那么如何自己创建一个 ContentProvider？要创建自己的 ContentProvider，需要遵循以下几步：

（1）创建一个继承了 ContentProvider 父类的子类。

（2）定义一个名为 CONTENT_URI，并且是 public static final 的 Uri 类型的类变量，用户必须为其指定唯一的字符串值，最好的方案是类的全名称，例如：

```
public static final Uri CONTENT_URI=Uri.parse( "content://com.google.
android.MyContentProvider");
```

（3）创建数据存储系统。大多数 ContentProvider 使用 Android 文件系统或 SQLite 数据库保存数据，但是也可以以任何想要的方式来存储。

（4）定义要返回给客户端的数据列名。如果正在使用 Android 数据库，则数据列的使用方式就和以往所熟悉的其他数据库一样。但是，必须为其定义一个名称为_id 的列，它用来表示每条记录的唯一性。

（5）如果要存储字节型数据，比如位图文件等，那保存该数据的数据列，其实是一个表示实际保存文件的 URI 字符串，客户端通过它来读取对应的文件数据，处理这种数据类型的 ContentProvider 需要实现一个名称为_data 的字段，_data 字段列出了该文件在 Android 文件系统上的精确路径。这个字段不仅是供客户端使用，而且也可以供 ContentResolver 使用。客户端可以调用 ContentResolver.openOutputStream()方法处理该 URI 指向的文件资源，如果是 ContentResolver 本身的话，由于其拥有的权限比客户端要高，所以它能直接访问该数据文件。

（6）声明 public static String 型的变量，用于指定要从游标处返回的数据列。

（7）查询返回一个 Cursor 类型的对象。所有执行写操作的方法如 insert()、update() 以及 delete()都将被监听。可以通过使用 ContentResover().notifyChange()方法通知监听器关于数据更新的信息。

（8）在 AndroidMenifest.xml 中使用标签来设置 ContentProvider。

（9）如果要处理的数据类型是一种比较新的类型，就必须先定义一个新的 MIME 类型，以供 ContentProvider.geType(url)返回。MIME 类型有两种形式:

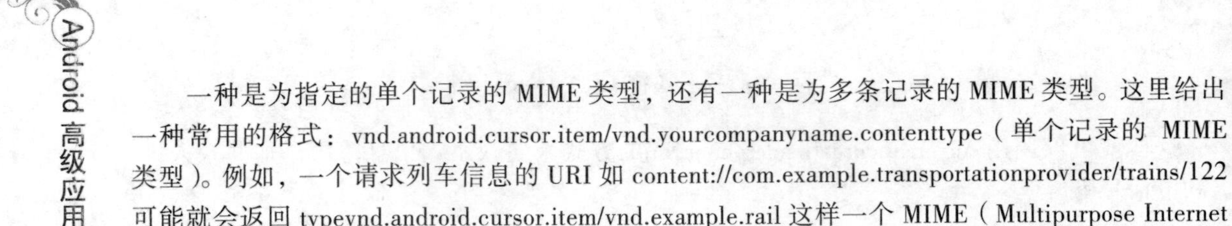

一种是为指定的单个记录的 MIME 类型，还有一种是为多条记录的 MIME 类型。这里给出一种常用的格式：vnd.android.cursor.item/vnd.yourcompanyname.contenttype（单个记录的 MIME 类型）。例如，一个请求列车信息的 URI 如 content://com.example.transportationprovider/trains/122 可能就会返回 typevnd.android.cursor.item/vnd.example.rail 这样一个 MIME（Multipurpose Internet Mail Extensions，多用途互联网邮件扩展）类型。

vnd.android.cursor.dir/vnd.yourcompanyname.contenttype （多个记录的 MIME 类型）。例如，一个请求所有列车信息的 URI 如 content://com.example.transportationprovider/trains 可能就会返回 vnd.android.cursor.dir/vnd.example.rail 这样一个 MIME 类型。

下列代码将创建一个 ContentProvider，它仅仅是存储用户名称并显示所有的用户名称（使用 SQLite 数据库存储这些数据）：

```
public class MyUsers {
    public static final String AUTHORITY="com.wissen.MyContentProvider";
    //BaseColumn 类中已经包含了 _id 字段
    public static final class User implements BaseColumns {
        public static final Uri CONTENT_URI=Uri.parse("content://com.wissen.
MyContentProvider");
        public static final String  USER_NAME="USER_NAME";    // 表数据列
    }
}
```

上面的类中定义了 ContentProvider 的 CONTENT_URI 以及数据列。下面将定义基于上面的类来定义实际的 ContentProvider 类：

```
package com.zigcloud.testApp;
public class MyContentProvider extends ContentProvider {
    private SQLiteDatabase sqlDB;
    private DatabaseHelper dbHelper;
    private static final String  DATABASE_NAME="Users.db";
    private static final int DATABASE_VERSION=1;
    private static final String TABLE_NAME="User";
    private static final String TAG="MyContentProvider";
    private static class DatabaseHelper extends SQLiteOpenHelper {
        DatabaseHelper(Context context) {
            super(context, DATABASE_NAME, null, DATABASE_VERSION);
        }
        public void onCreate(SQLiteDatabase db) {   //创建用于存储数据的表
            db.execSQL("Create table "+TABLE_NAME+"( _id INTEGER PRIMARY KEY
AUTOINCREMENT, USER_NAME TEXT);");
        }
        public void onUpgrade(SQLiteDatabase db,int oldVersion,int newVersion) {
            db.execSQL("DROP TABLE IF EXISTS "+TABLE_NAME);
            onCreate(db);
        }
    }
    public int delete(Uri uri, String s, String[] as) {
        return 0;
    }
    public String getType(Uri uri) {
        return null;
```

```
        }
        public Uri insert(Uri uri, ContentValues contentvalues) {
            sqlDB=dbHelper.getWritableDatabase();
            long rowId=sqlDB.insert(TABLE_NAME, "", contentvalues);
            if(rowId>0) {
                Uri rowUri=ContentUris.appendId(MyUsers.User.CONTENT_URI.
buildUpon(), rowId).build();
                getContext().getContentResolver().notifyChange(rowUri, null);
                return rowUri;
            }
            throw new SQLException("Failed to insert row into "+uri);
        }
        public boolean onCreate() {
            dbHelper=new DatabaseHelper(getContext());
            return (dbHelper==null) ? false : true;
        }
        public Cursor query(Uri uri, String[] projection, String selection,
String[] selectionArgs, String sortOrder) {
            SQLiteQueryBuilder qb=new SQLiteQueryBuilder();
            SQLiteDatabase db=dbHelper.getReadableDatabase();
            qb.setTables(TABLE_NAME);
            Cursor c=qb.query(db,projection,selection,null,null,null,sortOrder);
            c.setNotificationUri(getContext().getContentResolver(), uri);
            return c;
        }
        public int update(Uri uri,ContentValues contentvalues,String s,String[] as) {
            return 0;
        }
    }
```

一个名为 MyContentProvider 的 ContentProvider 创建完成了，它用于从 SQLite 数据库中添加和读取记录。ContentProvider 的入口需要在 AndroidManifest.xml 中配置，之后，才能使用这个定义好的 ContentProvider：

```
    public class MyContentDemo extends Activity {
        protected void onCreate(Bundle savedInstanceState) {
            super.onCreate(savedInstanceState);
            insertRecord("MyUser");
            displayRecords();
        }
        private void insertRecord(String userName) {
            ContentValues values=new ContentValues();
            values.put(MyUsers.User.USER_NAME, userName);
            getContentResolver().insert(MyUsers.User.CONTENT_URI, values);
        }
        private void displayRecords() {
            String columns[]=new String[]{MyUsers.User._ID, MyUsers.User.USER_NAME };
            Uri myUri=MyUsers.User.CONTENT_URI;
            Cursor cur=managedQuery(myUri, columns,null, null, null );
            if (cur.moveToFirst()) {
                String id=null;
```

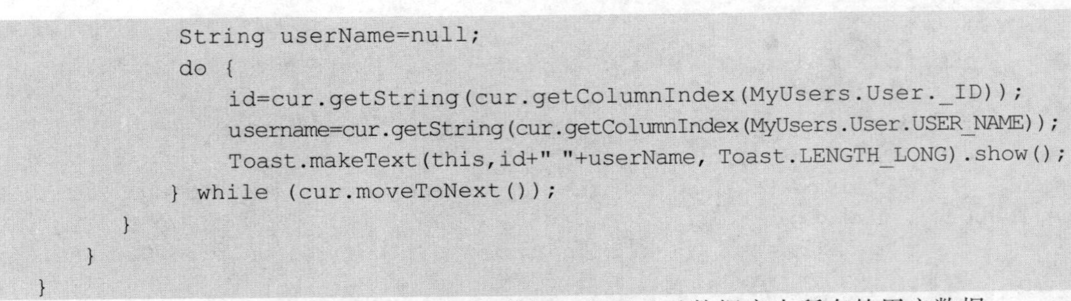

```
                String userName=null;
                do {
                    id=cur.getString(cur.getColumnIndex(MyUsers.User._ID));
                    username=cur.getString(cur.getColumnIndex(MyUsers.User.USER_NAME));
                    Toast.makeText(this,id+" "+userName, Toast.LENGTH_LONG).show();
                } while (cur.moveToNext());
            }
        }
    }
```

上面的类将先向数据库中添加一条用户数据，然后显示数据库中所有的用户数据。

ContentProvider 存储数据典型应用案例：短信监听

该程序通过监听 Uri(Content://sms)的数据改变即可监听用户短信的数据改变，并在监听器的onChange()方法里查询 Uri(content://sms/outbox)的数据，这样即可获取用户正在发送的短信。该程序的界面布局如图 2-8 所示，具体详细代码请参见配套光盘中的 CH02_4_1。

图 2-8　短信监听案例主界面

程序的界面布局代码如下：CH02\CH02_4\res\layout\ activity_main.xml。


```
<?xml version="1.0" encoding="utf-8"?>
<LinearLayout xmlns:android="http://schemas.android.com/apk/res/android"
  android:orientation="vertical"
  android:layout_width="fill_parent"
  android:layout_height="fill_parent">
<TextView
  android:layout_width="fill_parent"
  android:layout_height="wrap_content"
  android:text="@string/app_name"/>
</LinearLayout>
```

程序的 Java 代码如下：CH02\CH02_4\src\com\zigcloud\content\ MonitorSms.java。

```
public class MonitorSms extends Activity
{
  public void onCreate(Bundle savedInstanceState)
  {
     super.onCreate(savedInstanceState);
     setContentView(R.layout.activity_main);
     //为 content://sms 的数据改变注册监听器
     getContentResolver().registerContentObserver(
        Uri.parse("content://sms"), true,
        new SmsObserver(new Handler()));
  }
  private final class SmsObserver extends ContentObserver
  {  //提供自定义的 ContentObserver 监听器类
     public SmsObserver(Handler handler)
     {
        super(handler);
     }
     public void onChange(boolean selfChange)
     {   //查询发送箱中的短信(处于正在发送状态的短信放在发送箱)
```

```
            Cursor cursor=getContentResolver().query(Uri.parse("content://
sms/outbox"), null, null, null, null);
            //遍历查询得到的结果集，即可获取用户正在发送的短信
            while (cursor.moveToNext())
            {
               StringBuilder sb=new StringBuilder();
               //获取短信的发送地址
               sb.append("address=").append(cursor.getString(cursor.getColumnIndex
("address")));
               //获取短信的标题
               sb.append(";subject=").append(cursor.getString(cursor.getColumnIndex
("subject")));
               //获取短信的内容
               sb.append(";body=").append(cursor.getString(cursor.getColumnIndex
("body")));
               //获取短信的发送时间
               sb.append(";time=").append(cursor.getLong(cursor.getColumnIndex
("date")));
               System.out.println("Has Sent SMS:"+sb.toString());
            }
         }
      }
   }
```

该程序的运行效果如图 2-9 所示。

图 2-9　短信监听案例运行主界面

2.2.5　网络存储数据

前面介绍的几种存储都是将数据存储在本地设备上，除此之外，还有一种存储（获取）数据的方式，通过网络来实现数据的存储和获取。可以调用 WebService 返回的数据或是解析 HTTP（Hypertext Transfer Protocol，超文本传输协议）实现网络数据交互。具体需要熟悉 java.net.*，Android.net.*这两个包的内容，这里不再赘述，请大家参阅相关文档。下面是一个通过地区名称查询该地区的天气预报，以 POST 发送的方式发送请求到 webservicex.net 站点，访问 WebService.webservicex.net 站点上提供查询天气预报的服务。代码如下：

```
   public class MyAndroidWeatherActivity extends Activity {
      //定义需要获取的内容来源地址
      private static final String SERVER_URL="http://www.webservicex.net/
WeatherForecast.asmx/GetWeatherByPlaceName";

      public void onCreate(Bundle savedInstanceState) {
                                 /**第一次创建活动（activity ）时调用*/
         super.onCreate(savedInstanceState);
```

```
        setContentView(R.layout.main);
        HttpPost request=new HttpPost(SERVER_URL);
                                        //根据内容来源地址创建一个 HTTP 请求
        List<NameValuePair> params=new ArrayList<NameValuePair>();
                                        //添加一个变量
        //设置一个地区名称
        params.add(new BasicNameValuePair("PlaceName", "NewYork"));
                                            //添加必需的参数
         try { //设置参数的编码
            request.setEntity(new UrlEncodedFormEntity(params, HTTP.UTF_8));
            //发送请求并获取反馈
            HttpResponse httpResponse=new DefaultHttpClient().execute (request);
            if(httpResponse.getStatusLine().getStatusCode() != 404){
                                            //解析返回的内容
               String result=EntityUtils.toString(httpResponse.getEntity());
               System.out.println(result);
            }
        } catch(Exception e) {
            e.printStackTrace();
        }
    }
}
```

别忘记了在配置文件中设置访问网络权限：

```
<uses-permission android:name="android.permission.INTERNET" />
```

2.3　案例

2.3.1　基于 sharedPreferences 的参数配置文件

在 Android 应用程序中，通常会将软件常用的一些参数存储到本地，以便下次打开程序的时候不再去设置这些参数，所以会为每个应用程序添加一些配置文件。本案例 sharedPreferences 数据存储技术保持系统设置的定时刷机间隔、最小控制间隔和定时上传间隔等参数，该程序的界面布局如图 2-10 所示，其详细代码请参见配套光盘中的 CH02_5。

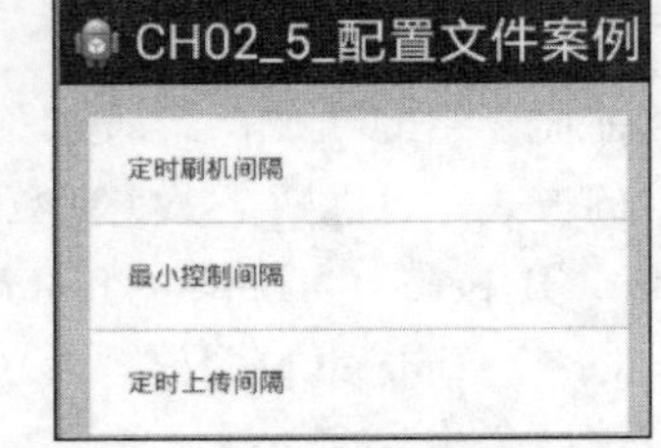

图 2-10　配置文件案例布局主界面

程序的界面布局代码如下：CH02\CH02_5\res\layout\activity_main.xml。

```
<RelativeLayout xmlns:android="http://schemas.android.com/apk/res/android"
    xmlns:tools="http://schemas.android.com/tools"
    android:layout_width="match_parent"
    android:layout_height="match_parent"
    android:paddingBottom="@dimen/activity_vertical_margin"
    android:paddingLeft="@dimen/activity_horizontal_margin"
    android:paddingRight="@dimen/activity_horizontal_margin"
```

```
android:paddingTop="@dimen/activity_vertical_margin"
android:background="#CCCCCC"
tools:context=".MainActivity" >
<LinearLayout
    android:layout_width="fill_parent"
    android:layout_height="fill_parent"
    android:layout_alignParentBottom="true"
    android:layout_alignParentRight="true"
    android:orientation="vertical" >
    <LinearLayout
        android:id="@+id/ll_refresh_interval"
        android:layout_width="match_parent"
        android:layout_height="wrap_content"
        android:layout_marginTop="1dp"
        android:background="#FFFFFF"
        android:gravity="center_vertical"
        android:orientation="horizontal"
        android:padding="20dp"
        android:clickable="true" >
        <TextView
            android:id="@+id/tv_refresh_interval"
            android:layout_width="wrap_content"
            android:layout_height="wrap_content"
            android:text="@string/refresh_interval" />
    </LinearLayout>
    <LinearLayout
        android:id="@+id/ll_control_interval"
        android:layout_width="match_parent"
        android:layout_height="wrap_content"
        android:layout_marginTop="1dp"
        android:background="#FFFFFF"
        android:gravity="center_vertical"
        android:orientation="horizontal"
        android:padding="20dp"
        android:clickable="true" >
        <TextView
          android:id="@+id/tv_control_interval"
          android:layout_width="wrap_content"
          android:layout_height="wrap_content"
          android:text="@string/control_interval" />
    </LinearLayout>
    <LinearLayout
        android:id="@+id/ll_upload_interval"
        android:layout_width="match_parent"
        android:layout_height="wrap_content"
        android:layout_marginTop="1dp"
```

```
            android:background="#FFFFFF"
            android:gravity="center_vertical"
            android:orientation="horizontal"
            android:padding="20dp"
            android:clickable="true">
            <TextView
                android:id="@+id/tv_upload_interval"
                android:layout_width="wrap_content"
                android:layout_height="wrap_content"
                android:text="@string/upload_interval" />
        </LinearLayout>
    </LinearLayout>
</RelativeLayout>
```

程序 Java 代码：CH02\CH02_5\src\com\example\demosharedpreferences\MainActivity.java。

```
public class MainActivity extends Activity implements  OnClickListener{
/***系统配置文件* */
    protected void onCreate(Bundle savedInstanceState) {
        super.onCreate(savedInstanceState);
        setContentView(R.layout.activity_main);
        findViewById(R.id.ll_refresh_interval).setOnClickListener(this);
        findViewById(R.id.ll_control_interval).setOnClickListener(this);
        findViewById(R.id.ll_upload_interval).setOnClickListener(this);
    }
    public boolean onCreateOptionsMenu(Menu menu) {
        getMenuInflater().inflate(R.menu.main, menu);
                                        //膨胀的菜单，将菜单项添加到行动条
        return true;
    }
    public void onClick(View v) {
        switch(v.getId()){
            case R.id.ll_refresh_interval: //设置定时刷新间隔
                setRereshInterval();
                break;
            case R.id.ll_control_interval: //设置最小控制间隔
                setControlInterval();
                break;
            case R.id.ll_upload_interval:  //设置自动上传间隔
                setUploadInterval();
                break;
        }
    }
    private void setRereshInterval(){/*** 设置定时刷新间隔* */
        //初始化对话框的 view，并设置编辑框的值
        View view=View.inflate(this, R.layout.view_config_refresh_interval, null);
        final EditText et_refresh_interval=(EditText)view.findViewById(R.id.
et_refresh_interval);
```

```
        et_refresh_interval.setText(String.valueOf(ApplicationConfig.
getInstance(this).getRefreshInterval()));
        AlertDialog.Builder builder=new Builder(this);   //创建对话框并显示
        AlertDialog alertDialog=builder.setTitle(R.string.refresh_interval).
setView(view).setPositiveButton(R.string.ok, new DialogInterface.OnClickListener() {
            public void onClick(DialogInterface dialog, int which) {
                ApplicationConfig.getInstance(getParent()).setRefreshInterval
(Long.parseLong(et_refresh_interval.getText().toString()));
                ApplicationConfig.getInstance(getParent()).save();
            }
        })
        .setNegativeButton(R.string.cancel,new DialogInterface. OnClickLis
tener() {
            public void onClick(DialogInterface dialog, int which) {
            }
        })
        .create();
        alertDialog.show();
    }
    private void setControlInterval(){/*** 设置最小控制间隔* */
        //初始化对话框的 view，并设置编辑框的值
        View view=View.inflate(this, R.layout.view_config_control_interval, null);
        final EditText et_control_interval=(EditText)view.findViewById(R.id.
et_control_interval);
        et_control_interval.setText(String.valueOf(ApplicationConfig.
getInstance(this).getControlInterval()));
        AlertDialog.Builder builder=new Builder(this);
                                            //创建对话框并显示
        AlertDialog alertDialog=builder.setTitle(R.string.control_interval).
setView(view).setPositiveButton(R.string.ok,new DialogInterface. OnClickListener() {
            public void onClick(DialogInterface dialog, int which) {
              ApplicationConfig.getInstance(getParent()).setControlInterval
(Long.parseLong(et_control_interval.getText().toString()));
                    ApplicationConfig.getInstance(getParent()).save();
                }
            })
            .setNegativeButton(R.string.cancel,new DialogInterface. OnClickListener() {
            public void onClick(DialogInterface dialog, int which) {
            }
        })
        .create();
        alertDialog.show();
    }
    private void setUploadInterval(){/*** 设置自动上传间隔* */
        //初始化对话框的 view，并设置编辑框的值
        View view=View.inflate(this,R.layout.view_config_upload_interval, null);
```

```
        final EditText et_upload_interval=(EditText)view.findViewById(R.id.
et_upload_interval);
        et_upload_interval.setText(String.valueOf(ApplicationConfig.getInstance(this).
getUploadInterval()));
        AlertDialog.Builder builder=new Builder(this);
                                        //创建对话框并显示
        AlertDialog alertDialog=builder.setTitle(R.string.upload_interval)
.setView(view).setPositiveButton(R.string.ok,new DialogInterface.
OnClickListener() {
            public void onClick(DialogInterface dialog, int which) {
                ApplicationConfig.getInstance(getParent()).setUploadInterval
(Long.parseLong(et_upload_interval.getText().toString()));
                ApplicationConfig.getInstance(getParent()).save();
            }
        })
        .setNegativeButton(R.string.cancel, new DialogInterface. OnClickLi
stener() {
            public void onClick(DialogInterface dialog, int which) {
            }
        })
        .create();
        alertDialog.show();
    }
}
```

程序清单：CH02\ CH02_5\com\example\demosharepreference\applicaiton \ApplicationConfig.java。

```
public long getUploadInterval() {    /***获取定时上传间隔**/
        long uploadInterval=sharedPreferences.getLong("uploadInterval", 3000);
        return uploadInterval;
    }
    public void setUploadInterval(long autobackStandardModel) {
    /***设置定时上传间隔**/
        editor.putLong("uploadInterval", autobackStandardModel);
    }
    public long getRefreshInterval() { /***获取定时刷机间隔**/
        long refreshInterval=sharedPreferences.getLong("refreshInterval", 10000);
        return refreshInterval;
    }
    public void setRefreshInterval(long refreshInterval) {
    /***设置定时刷机间隔**/
        editor.putLong("refreshInterval", refreshInterval);
    }
    public long getControlInterval() { /***获取最小控制间隔**/
        long controlInterval=sharedPreferences.getLong("controlInterval", 1000);
        return controlInterval;
    }
    public void setControlInterval(long controlInterval) {
```

```
    /***设置最小控制间隔**/
        editor.putLong("controlInterval", controlInterval);
    }
    public static ApplicationConfig getInstance(Context context){
    /***获取应用程序配置文件**/
        if(applicationConfig==null){
            applicationConfig=new ApplicationConfig(context);
        }
        return applicationConfig;
    }
    private ApplicationConfig(Context context){ /***私有构造函数**/
      this.sharedPreferences = context.getSharedPreferences("preferences",
Context.MODE_PRIVATE);
        this.editor = sharedPreferences.edit();
    }
    public boolean save(){/***保存**/
        return editor.commit();
    }
}
```

该程序的运行效果如图 2-11 所示。

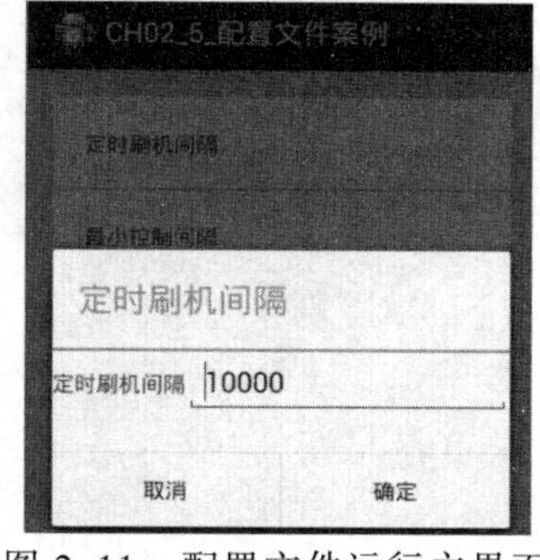

图 2-11　配置文件运行主界面

2.3.2　基于 SQLite 数据库的设备状态信息显示

跟 web 一些程序类似，在 Android 程序中通常会遇到一些数据需要存储、查询和显示，这时候就需要使用到嵌入式平台上最流行的数据库 SQLite 存储数据，本案例利用 SQLite 数据库技术，显示物联网智能家居实训系统中设备（如风扇、窗帘、壁灯和窗灯等）状态。该案例利用 SQLite 数据库存储设备信息，在程序启动的时候读取数据库中的数据呈现在界面上，该程序的界面布局如图 2-12 所示，。其详细代码请参见配套光盘中的 CH02_6。

程序的界面布局代码如下：CH02\CH02_6\res\layout\activity_main.xml。

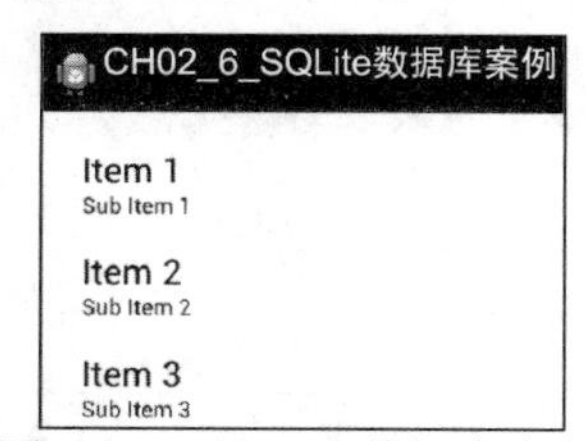

图 2-12　SQLite 数据库案例布局主界面

```
    <RelativeLayout xmlns:android="http://schemas.
android.com/apk/res/android"
        xmlns:tools="http://schemas.android.com/ tools"
        android:layout_width="match_parent"
        android:layout_height="match_parent"

android:paddingBottom="@dimen/activity_vertical_margin"
        android:paddingLeft="@dimen/activity_horizontal_margin"
        android:paddingRight="@dimen/activity_horizontal_margin"
        android:paddingTop="@dimen/activity_vertical_margin"
        tools:context=".MainActivity" >
        <GridView
            android:id="@+id/gridView1"
```

```
        android:layout_width="match_parent"
        android:layout_height="wrap_content"
        android:layout_alignParentLeft="true"
        android:layout_alignParentTop="true"
        android:numColumns="1" >
    </GridView>
</RelativeLayout>
```

程序清单 CH02\CH02_6\res\layout\ activity_equipment_list_item.xml。

```
<?xml version="1.0" encoding="utf-8"?>
<RelativeLayout xmlns:android="http://schemas.android.com/apk/res/android"
    android:id="@+id/RelativeLayout1"
    android:layout_width="match_parent"
    android:layout_height="wrap_content" >
    <ImageView
        android:id="@+id/img_image"
        android:layout_width="wrap_content"
        android:layout_height="wrap_content"
        android:layout_alignParentLeft="true"
        android:layout_alignParentTop="true"
        android:src="@drawable/ic_launcher" />
    <TextView
        android:id="@+id/tv_datavalue"
        android:layout_width="wrap_content"
        android:layout_height="wrap_content"
        android:layout_alignBottom="@+id/img_image"
        android:layout_alignLeft="@+id/tv_name"
        android:text="TextView" />
    <TextView
        android:id="@+id/tv_area"
        android:layout_width="wrap_content"
        android:layout_height="wrap_content"
        android:layout_above="@+id/tv_datavalue"
        android:layout_alignParentRight="true"
        android:layout_alignParentTop="true"
        android:text="TextView" />
    <TextView
        android:id="@+id/tv_name"
        android:layout_width="wrap_content"
        android:layout_height="wrap_content"
        android:layout_alignParentTop="true"
        android:layout_toRightOf="@+id/img_image"
        android:layout_marginLeft="15dp"
        android:text="TextView" />
</RelativeLayout>
```

程序的 Java 代码：CH02\CH02_6\src\ com\example\demosqlite\ MainActivity.java。

```
public class MainActivity extends Activity {/*** 主页面* */
   private EquipmentDAO mEquipmentDAO=null;    //设备业务类
```

```
    private GridView mGridView=null;             //显示设备的 GridView
    private SimpleAdapter mSimpleAdapter=null; //设备列表适配器
    private List<HashMap<String, Object>> mHashMapList=null;
                                                 //设备信息数据源
    protected void onCreate(Bundle savedInstanceState) {
        super.onCreate(savedInstanceState);
        setContentView(R.layout.activity_main);
        initial();                               //初始化
        loadData();                              //加载数据
    }
    public boolean onCreateOptionsMenu(Menu menu) {
        getMenuInflater().inflate(R.menu.main, menu);
                                             //膨胀的菜单，将菜单项添加到行动条
        return true;
    }
    private void initial(){   /*** 初始化* */
        mEquipmentDAO=new EquipmentDAO(this);//初始化设备业务类
        //初始化设备信息数据源
        mHashMapList=new ArrayList<HashMap<String,Object>>();
        mGridView=(GridView)findViewById(R.id.gridView1);
                                             //初始化 GridView
        //初始化适配器
        mSimpleAdapter=new SimpleAdapter(this,mHashMapList,R.layout. activity_
equipment_list_item,new String[] {"ItemImage","ItemNodeName","ItemDataValue"},
new int[] {R.id.img_image,R.id.tv_name,R.id.tv_datavalue});
        mGridView.setAdapter(mSimpleAdapter);
    }
    private void loadData(){/***加载 SQLite 数据库中的数据**/
        List<EquipmentEntity> equipmentList=mEquipmentDAO.getAll();
        if(equipmentList!=null){
            EquipmentEntity equipmentEntity=null;
            HashMap<String, Object> item=null;
            for(int i=0;i<equipmentList.size();i++){
                equipmentEntity=equipmentList.get(i);
                if(equipmentEntity!=null){
                    item=new HashMap<String, Object>();
                    item.put("ItemImage", R.drawable.ic_launcher);
                    item.put("ItemNodeName", equipmentEntity.getName());
                    item.put("ItemDataValue",equipmentEntity.getLastUpdateTime());
                    item.put("ItemArea", "");
                    mHashMapList.add(item);
                }
            }
            mSimpleAdapter.notifyDataSetChanged();
        }
    }
}
```

程序清单 CH02\CH02_6\src\com\example\biz\EquipmentDAO.java。

```
public class EquipmentDAO {/*** 设备业务类* */
    private DBOpenHelper helper;
    public EquipmentDAO(Context context){
        helper=new DBOpenHelper(context);
    }
    public List<EquipmentEntity> getAll(){/***获取所有设备**/
        ArrayList<EquipmentEntity> equpmentList=new ArrayList<EquipmentEntity>();
        SQLiteDatabase db=helper.getReadableDatabase();
        String sql="select [id],[name],[nodeID],[nodeTypeID],[lastUpdateTime]
from [equipment]";
        Cursor c=db.rawQuery(sql, null);
        while(c.moveToNext()){
          equpmentList.add(new EquipmentEntity(c.getInt(0), c.getString(1),
c.getString(2), c.getString(3), c.getString(4)));
        }
        c.close();
        db.close();
        return equpmentList;
    }
}
```

程序清单 CH02\CH02_6\src\ com\example\db\DBOpenHelper.java。

```
public class DBOpenHelper extends SQLiteOpenHelper {/*** 数据库助手类* */
    private static final String DATABASENAME="data.db";  /***Database name**/
    private static final int DATABASEVERSION=2; /*** Database Version* */
    public DBOpenHelper(Context context) {
        super(context, DATABASENAME, null, DATABASEVERSION);
    }
    public void onCreate(SQLiteDatabase db) {
        db.execSQL("CREATE TABLE [equipment]("
                + "[id] integer not null primary key autoincrement,"
                + "[name] varchar(255),"
                + "[nodeID] varchar(255),"
                + "[nodeTypeID] varchar(255),"
                + "[lastUpdateTime] varchar(255))");
        db.execSQL("insert into [equipment]([name],[nodeID],[nodeTypeID],
[lastUpdateTime]) values('风扇 01','01','0001','2014/1/14 12:00:20')");
        db.execSQL("insert into [equipment]([name],[nodeID],[nodeTypeID],
[lastUpdateTime]) values('窗帘 02','02','0002','2014/1/14 12:12:06')");
        db.execSQL("insert into [equipment]([name],[nodeID],[nodeTypeID],
[lastUpdateTime]) values('壁灯 03','03','0003','2014/1/14 12:02:08')");
        db.execSQL("insert into [equipment]([name],[nodeID],[nodeTypeID],
[lastUpdateTime]) values('窗灯 04','04','0004','2014/1/14 12:08:50')");
    }
    public void onUpgrade(SQLiteDatabase db, int oldVersion, int newVersion) {
        db.execSQL("DROP TABLE IF EXISTS [equipment]");
        onCreate(db);
    }
}
```

程序清单 CH02\CH02_6\src\ com\example\entity\EquipmentEntity.java。

```
package com.example.entity;
public class EquipmentEntity {/*** 设备实体类* */
    private int id; /*** 编号* */
    private String name; /*** 结点名称* */
    private String nodeID; /*** 结点编号* */
    private String nodeTypeID;   /*** 结点类型* */
    private String lastUpdateTime; /*** 最后更新时间* */
    public EquipmentEntity(){
    }
    public EquipmentEntity(int id,String name,String nodeID,String nodeTypeID,
String lastUpdateTime){
        this.id=id;
        this.name=name;
        this.nodeID=nodeID;
        this.nodeTypeID=nodeTypeID;
        this.lastUpdateTime=lastUpdateTime;
    }
    public int getId() {
        return id;
    }
    public void setId(int id) {
        this.id=id;
    }
    public String getName() {
        return name;
    }
    public void setName(String name) {
        this.name=name;
    }
    public String getNodeID() {
        return nodeID;
    }
    public void setNodeID(String nodeID) {
        this.nodeID=nodeID;
    }
    public String getNodeTypeID() {
        return nodeTypeID;
    }
    public void setNodeTypeID(String nodeTypeID) {
        this.nodeTypeID=nodeTypeID;
    }
    public String getLastUpdateTime() {
        return lastUpdateTime;
    }
    public void setLastUpdateTime(String lastUpdate Time) {
        this.lastUpdateTime=lastUpdateTime;
    }
}
```

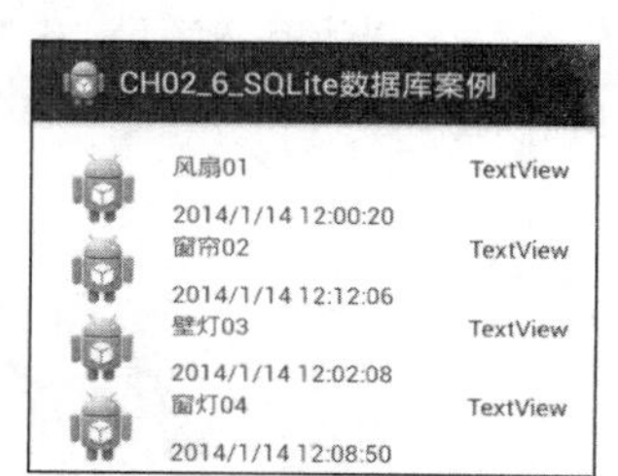

图 2-13　SQLite 数据库案例运行主界面

该程序的运行效果如图 2-13 所示。

2.4 知识扩展

数据共享（ContentProvider）

（1）ContentProvider 的基本概念。

（2）ConentProvider 是一种特殊的存储数据的类型，它提供了一套标准的接口来获取、操作数据。

（3）ContentReslover。

（4）URI 与 URL。

（5）ContentReslover 主要接口。

（6）ContentProvider 的使用方法。

（7）ContentProvider 典型应用案例。

2.5 本章小结

本章主要介绍了 Android 的五种不同的数据存储方式，SharedPreferences 存储数据、文件存储数据、SQLite 数据库存储数据、ContentProvider 存储数据、网络存储数据的特点及其用法。

2.6 强化练习

1．填空题

（1）Android 的五种不同的数据存储方式为（　　）、（　　）、（　　）、（　　）、（　　）。

（2）SharedPreferences 提供了 Android 平台常规的（　　）、（　　）、（　　）、（　　）、（　　）数据数类的保存。

（3）SharedPreferences 是基于 xml 文件存储（　　）数据。

（4）Activity 提供了（　　）方法可以用于把数据输出到文件中。

（5）SQLite 基本上符合（　　）标准，和其他的主要 SQL 数据库没什么区别。

（6）可以调用（　　）返回的数据或是解析（　　）实现网络数据交互。

2．问答题、操作题或编程题

（1）简述实现 SharedPreferences 存储的步骤。

（2）SQLite 是轻量级嵌入式数据库引擎，其特点有哪些？

（3）ContentProvider 最大的特点是什么？

第3章 Android 网络应用

学习目标：

- 理解 Android 常用网络的几种基本概念；
- 掌握 HTTP 通信的实现；
- 掌握 Socket 通信的实现。

3.1 学习导入

所谓无线网络，即采用无线传输媒介如无线电波、红外线等网络。既包括允许用户建立远距离无线连接的全球语音和数据网络，也包括为近距离无线连接进行优化的红外线技术以及射频技术。从无线网络的名字上就可以推断出它与有线网络的主要区别就是通过无线电技术取代的网线，二者互为备份。下面就来介绍一下无线网络的基础知识。

3.1.1 窄带广域网

HSCSD（高速线路交换数据）是为无线用户提供 38.3 kbit/s 速率传输的无线数据传输方式，它的速度比 GSM 通信标准的标准数据速率快 4 倍，可以和使用固定电话调制解调器的用户相比。当前，GSM 网络单个信道在每个时隙只能支持 1 个用户，而 HSCSD 通过允许 1 个用户在同一时间同时访问多个信道来大幅改进数据访问速率。但美中不足的是，这会导致用户成本的增加。假设 1 个标准的数据传输速率是 14 400 bit/s，使用具有 4 个时隙的 HSCSD 将使数据访问速率达到 57.6 kbit/s。

GPRS（多时隙通用分组无线业务）是一种很容易与 IP 接口的分组交换业务，其速率可达 9.6～14.4 kbit/s，甚至能达到 115 kbit/s，并且能够传送语音和数据。该技术是当前提高 Internet 接入速度的热门技术，而且还有可能被应用在广域网中。GPRS 又被认为是 GSM 第 2 阶段增强(GSM Phase2 +)接入技术。GPRS 虽是 GSM 上的分组数据传输标准，但也可和 IS-136 标准结合使用。

GPRS 是 GSM 一项承载业务，提高并简化了无线数据接入分组网络的方式，分组数据可直接在 GSM 基站和其他分组网之间传输。它具有接入时间短、速率高的特点。由于它是分组方式的，因此可以按字节数来计费，这些和传统的拨号接入时间长、按电路持续时间计费明显不同。同时，GPRS 网是 GSM 上的分组网，它实际上又是 Internet 的一个子网。在 GPRS 的支持下，GSM 可以提供：E-mail、网页浏览、增强的短消息业务、即时的无线图像传送、寻像业务、文本和住处共享、监视、Voice over Internet、广播业务。由于它采用的是分组技术，与传统的无线电路业务在实施上有完全不同的特点。

GPRS 网络同时支持 IPv4 和 IPv6，是通向第三代移动通信网络的重要一步。它适合于突发性 Internet/ Intranet 业务，并能提供点到点的承载业务以及完成短消息业务的传送。预计在将来，它也能提供单点到多点的业务。更重要的是 GPRS 具有有限的 QoS 支持，因为它可以由相关参数来指定业务的继承性、可靠性、延时、流量。

CDPD（蜂窝数字分组数据）采用分组数据方式，是目前公认的最佳无线公共网络数据通信规程。它是建立在 TCP/IP 基础上的一种开放系统结构，将开放式接口、高传输速度、用户单元确定、空中链路加密、空中数据加密、压缩数据纠错及重发和世界标准的 IP 寻址模式无线接入有机地结合在一起，提供同层网络的无缝连接、多协议网络服务。

3.1.2 宽带广域网

目前随着 3G 网络的不断普及与成熟，3G 标准也分为三种不同的标准。下面就来介绍一下目前存在的 3 种 3G 标准：它们分别是 W-CDMA（欧洲版）、CDMA2000（美国版）和 TD-SCDMA（中国版）。

WCDMA 为 Wideband CDMA，又称 CDMA Direct Spread，意为宽频分码多重存取，这是基于 GSM 网络发展出来的 3G 技术规范，是欧洲提出的宽带 CDMA 技术，它与日本提出的宽带 CDMA 技术基本相同，目前正在进一步融合。W-CDMA 的支持者主要是以 GSM 系统为主的欧洲厂商，日本公司也或多或少参与其中，包括欧美的爱立信、阿尔卡特、诺基亚、朗讯、北电，以及日本的 NTT、富士通、夏普等厂商。该标准提出了 GSM（2G）-GPRS- EDGE-WCDMA（3G）的演进策略。这套系统能够架设在现有的 GSM 网络上，对于系统提供商而言可以较轻易地过渡。预计在 GSM 系统相当普及的亚洲，对这套新技术的接受度会相当高。因此 W-CDMA 具有先天的市场优势。

CDMA2000 是由窄带 CDMA（CDMA IS95）技术发展而来的宽带 CDMA 技术，又称 CDMA Multi-Carrier，它是由美国高通北美公司为主导提出，摩托罗拉、Lucent 和后来加入的韩国三星都有参与，韩国现在成为该标准的主导者。这套系统是从窄频 CDMAOne 数字标准衍生出来的，可以从原有的 CDMAOne 结构直接升级到 3G，建设成本低廉。但目前使用 CDMA 的地区只有日本、韩国和北美，所以 CDMA2000 的支持者不如 W-CDMA 多。不过 CDMA2000 的研发技术却是目前各标准中进度最快的，许多 3G 手持设备已经率先面世。该标准提出了从 CDMA IS95（2G）-CDMA20001x-CDMA20003x（3G）的演进策略。CDMA20001x 被称为 2.5 代移动通信技术。CDMA20003x 与 CDMA20001x 的主要区别在于应用了多路载波技术，通过采用三载波使带宽提高。目前中国电信正在采用这一方案向 3G 过渡，并已建成了 CDMA IS95 网络。

TD-SCDMA 全称为 Time Division - Synchronous CDMA（时分同步 CDMA），该标准是由我国内地独自制定的 3G 标准，1999 年 6 月 29 日，中国原邮电部电信科学技术研究院（大唐电信）向 ITU 提出，但技术发明始于西门子公司，TD-SCDMA 具有辐射低的特点，被誉为绿色 3G。该标准将智能无线、同步 CDMA 和软件无线电等当今国际领先技术融于其中，在频谱利用率、对业务支持具有灵活性、频率灵活性及成本等方面具有独特优势。另外，由于中国内地庞大的市场，该标准受到各大主要电信设备厂商的重视，全球一半以上的设备厂商都宣布可以支持 TD—SCDMA 标准。该标准提出不经过 2.5 代的中间环节，直接向 3G 过渡，非常适用于 GSM 系统向 3G 升级。军用通信网也是 TD-SCDMA 的核心任务。

WiMAX 的全名是微波存取全球互通（Worldwide Interoperability for Microwave Access），又称为 801·26 无线城域网，是又一种为企业和家庭用户提供“最后一英里”的宽带无线连接方案。将此技术与需要授权或免授权的微波设备相结合之后，由于成本较低，将扩大宽带无线市场，改善企业与服务供应商的认知度。2007 年 10 月 19 日，在国际电信联盟在日内瓦举行的无线通信全体会议上，经过多数国家投票通过，WiMAX 正式被批准成为继 WCDMA、CDMA2000 和 TD-SCDMA 之后的第四个全球 3G 标准。

3.1.3 HTTP 通信

HTTP（Hyper Text Transfer Protocol，超文本传输协议）用于传送 WWW 方式的数据。HTTP 采用了请求/响应模型。客户端向服务器发送一个请求，请求头包含了请求的方法、URI、协议版本，以及包含请求修饰符、客户信息和内容的类似于 MIME 的消息结构。服务器以一个状态行作为响应，响应的内容包括消息协议的版本、成功或者错误编码，还包括服务器信息、实体元信息以及可能的实体内容。

Google 以网络搜索引擎著称，自然而然也会使 Android SDK 拥有强大的 HTTP 访问能力。在 Android SDK 中，Google 集成了 Apache 的 HttpClient 模块。要注意的是，这里的 Apache HttpClient 模块是 HttpClient4.0（org.apache.http.*），而不是 Jakarta Commons HttpClient 3.x（org.apache.commons.httpclient.*）。

3.1.4 Socket 通信

在 HTTP 通信中客户端每次发送的请求都需要服务器响应，在请求结束后会主动释放连接。将从连接的建立到关闭连接的过程称为“一次连接”。当需要保持客户端在线的状态就要不停地向服务器端发送连接请求。若服务器长时间无法收到客户端的请求，则认为客户端下线，若客户端长时间无法收到服务器的回复，则认为网络已经断开。在大多数情况下，需要服务器主动向客户端发送数据，保持客户端与服务器数据的同步，若通过 HTTP 建立连接，则服务器需要等到客户端发送一次请求后才能将数据回复给客户端，因此客户端定时向服务器发送连接请求不仅可以保持在线，同时也是在“询问”服务器是否有新的数据，如果有新数据就传送给客户端。但是当需要多人同时在线联网时，HTTP 连接就不能很好得满足需要，所以就需要 Socket 通信。

3.1.5 Wi-Fi

Wi-Fi 原先是无线保真的缩写，Wi-Fi 的英文全称为 Wireless Fidelity，在无线局域网的范畴是指“无线相容性认证”，实质上是一种商业认证，同时也是一种无线联网的技术，以前通过网线连接计算机，而现在则是通过无线电波来联网；常见的就是一个无线路由器，那么在这个无线路由器的电波覆盖的有效范围都可以采用 Wi-Fi 连接方式进行联网，如果无线路由器连接了一条 ADSL 线路或者别的上网线路，则又被称为“热点”。Wi-Fi 无线通信技术的品牌目前由 Wi-Fi 联盟（Wi-Fi Alliance）所持有。

Wi-Fi 的频段在世界范围内是不需要任何电信运营执照的免费频段，因此 WLAN 无线设备提供给了一个世界范围内可用的，费用及其低廉且数据带宽极高的无线空中接口。用户可以在 Wi-Fi 覆盖的区域内快速浏览网页等。同时，Wi-Fi 又有其突出的优势：

- 无线电波的覆盖范围广，基于蓝牙技术的电波覆盖范围非常小，半径只有 15 m 左右，而 Wi-Fi 的半径则可达 100 m，办公室自不用说，就是在整栋大楼中也可使用。最近，由 Vivato 公司推出的一款新型交换机。据悉，该款产品能够把目前 Wi-Fi 无线网络 100 m 的通信距离扩大到 6.5 km。
- 虽然由 Wi-Fi 技术传输的无线通信质量不是很好，数据安全性能比蓝牙差一些，传输质量也有待改进，但传输速度非常快，可以达到 54 Mbit/s，符合个人和社会信息化的需求。
- 厂商进入该领域的门槛比较低。厂商只要在机场、车站、咖啡店、图书馆等人员较密集的地方设置"热点"，并通过高速线路将因特网接入上述场所。这样，由于"热点"所发射出的电波可以达到距接入点半径 100 m 的地方，用户只要将支持无线 LAN 的笔记本式计算机或 PDA 拿到该区域内，即可高速接入因特网。也就是说，厂商不用耗费资金进行网络布线接入，从而节省了大量的成本。

在 Android 系统中，提供了 android.net.wifi 包供操作，在这个包中列了许多的信息，这里只介绍几个比较常用的操作，主要包括四个主要的类 ScanResult、wifiConfiguration、WifiInfo、WifiManager。

（1）ScanResult，主要是通过 Wi-Fi 硬件的扫描来获取一些周边的 Wi-Fi 热点的信息。在进行 WI-FI 搜索的时候，一般会搜到这些信息，首先是接入点名字、接入点信息的强弱、还有接入点使用的安全模式，是 WPA、WPE。打开这个类，可以看到以下几个信息：BSSID 接入点的地址，这里主要是指小范围内几个无线设备相连接所获取的地址，比如说两台笔记本式计算机通过无线网卡进行连接，双方的无线网卡分配的地址；SSID 网络的名字，当搜索一个网络时，就是靠这个来区分每个不同的网络接入点；Capabilities 网络接入的性能，这里主要是来判断网络的加密方式等；Frequency 频率，每一个频道交互的 MHz 数；Level 等级，主要用来判断网络连接的优先数。

（2）wifiConfiguration 在连通一个 Wi-Fi 接入点的时候，需要获取到的一些信息。大家可以跟有线设备进行对比一下。这里的数据相对来说比较复杂。六个子类 WifiConfiguration.AuthAlgorthm 用来判断加密方法。WifiConfiguration.GroupCipher 获取使用 GroupCipher 的方法进行加密。WifiConfiguration.KeyMgmt 获取使用 KeyMgmt 进行。WifiConfiguration.PairwiseCipher 获取使用 WPA 方式的加密。WifiConfiguration.Protocol 获取使用协议的加密。wifiConfiguration.Status 获取当前网络的状态。对于上述加密感兴趣的读者，可以在网上搜索相关的内容。

（3）WifiInfo 在 Wi-Fi 已经连通了以后，可以通过这个类获得一些已经连通的 Wi-Fi 连接的信息获取当前链接的信息，这里简单介绍一下这里的方法：

- getBSSID()：获取 BSSID。
- getDetailedStateOf()：获取客户端的连通性。
- getHiddenSSID()：获得 SSID 是否被隐藏。
- getIpAddress()：获取 IP 地址。
- getLinkSpeed()：获得连接的速度。
- getMacAddress()：获得 Mac 地址。
- getRssi()：获得 802.11n 网络的信号。
- getSSID()：获得 SSID。

- getSupplicanState()：返回具体客户端状态的信息。

（4）wifiManager 这个不用说，就是用来管理的 Wi-Fi 连接，这里已经定义好了一些，可以供使用。这里相对复杂，里面的内容比较多，但是通过字面意思，还是可以获得很多相关的信息。这个类里面预先定义了许多常量，可以直接使用，不用再次创建。这里还是简单介绍一下这里的方法：

- addNetwork(WifiConfiguration config)：通过获取到的网络的连接状态信息，来添加网络。
- calculateSignalLevel(int rssi , int numLevels)：计算信号的等级。
- compareSignalLevel(int rssiA, int rssiB)：对比连接 A 和连接 B。
- createWifiLock(int lockType, String tag)：创建一个 Wi-Fi 锁，锁定当前的 Wi-Fi 连接。
- disableNetwork(int netId)：让一个网络连接失效。
- disconnect()：断开连接。
- enableNetwork(int netId, Boolean disableOthers)：连接一个连接。
- getConfiguredNetworks()：获取网络连接的状态。
- getConnectionInfo()：获取当前连接的信息。
- getDhcpInfo()：获取 DHCP 的信息。
- getScanResulats()：获取扫描测试的结果。
- getWifiState()：获取一个 Wi-Fi 接入点是否有效。
- isWifiEnabled()：判断一个 Wi-Fi 连接是否有效。
- pingSupplicant()：ping 一个连接，判断是否能连通。
- ressociate()：即便连接没有准备好，也要连通。
- reconnect()：如果连接准备好了，连通。
- removeNetwork()：移除某一个网络。
- saveConfiguration()：保留一个配置信息。
- setWifiEnabled()：让一个连接有效。
- startScan()：开始扫描。
- updateNetwork(WifiConfiguration config)：更新一个网络连接的信息。

此外 wifiManaer 还提供了一个内部的子类，也就是 wifiManagerLock，wifiManagerLock 的作用是这样的，在普通状态下，如果 Wi-Fi 的状态处于闲置，那么网络的连通，将会暂时中断。但是如果把当前的网络状态锁上，那么 Wi-Fi 连通将会保持在一定状态，当然接触锁定之后，就会恢复常态。

Wi-Fi 典型应用案例：附近热点 Wi-Fi 搜索

该案例通过 Wi-Fi 来搜索附近的 Wi-Fi 可用热点。由于模拟器中不支持 Wi-Fi，所以程序的运行需要通过真机进行测试。首先看一下布局文件，在布局文件中，在线性布局中嵌套了两个不同的线性布局方式，其源代码清单如下所示，其运行界面如图 3-1 所示。

图 3-1 Wi-Fi 实训案例界面

主布局代码清单：CH03\CH03_1\res\layout\ main.xml。

```
<?xml version="1.0" encoding="utf-8"?>
<LinearLayout xmlns:android="http://schemas.android.com/apk/res/android"
```

```
    android:orientation="vertical"
    android:layout_width="fill_parent"
    android:layout_height="fill_parent" >
<LinearLayout android:id="@+id/btnSwitch"
     android:orientation="horizontal"
     android:layout_width="fill_parent"
     android:layout_height="wrap_content" >
     <Button android:id="@+id/btnOpenWIFI"
        android:layout_height="wrap_content"
        android:layout_width="wrap_content"
        android:text="打开 WIFI"/>
     <Button android:id="@+id/btnClose"
        android:layout_width="wrap_content"
        android:layout_height="wrap_content"
        android:text="关闭 WIFI"/>
</LinearLayout>
<LinearLayout android:id="@+id/listlayout"
     android:orientation="vertical"
     android:layout_width="fill_parent"
     android:layout_height="wrap_content" >
     <Button android:id="@+id/search"
     android:layout_width="fill_parent"
     android:layout_height="wrap_content"
     android:text="search"/>
     <TextView android:id="@+id/textV"
     android:layout_width="fill_parent"
     android:layout_height="wrap_content"
     android:text="List"/>
     <ListView android:id="@+id/android:list"
     android:layout_width="fill_parent"
     android:layout_height="wrap_content"
     android:drawSelectorOnTop="false"
     android:scrollbars="vertical"/>
</LinearLayout>
</LinearLayout>
```

主程序代码通过 getSystemService(WIFI_SERVICE)获 Wi-Fi 服务，通过 wifimanager.getScanResults()获取有效可用的 Wi-Fi 热点。通过两个按钮控制 Wi-Fi 功能的打开与关闭并同时使用一个按钮进行搜索。

源代码清单：CH03\CH03_1\src\sziit\ex\ch03\CH03_1.java。

```
public class WIFITest extends Activity {
   WifiManager wifimanager;
   List<ScanResult> wifilist;
   List<String> list;
   public void onCreate(Bundle savedInstanceState) {
      /** 第一次创建活动（activity ）时调用*/
      super.onCreate(savedInstanceState);
      setContentView(R.layout.main);
      final Button btnOpen=(Button)findViewById(R.id.btnOpenWIFI);
```

```
        Button btnClose=(Button)findViewById(R.id.btnClose);
        Button btnSearch=(Button)findViewById(R.id.search);
        final ListView listview=(ListView)findViewById(android.R.id.list);
        wifimanager=(WifiManager)getSystemService(WIFI_SERVICE);
                                          //获取 WifiManager 对象
        btnOpen.setOnClickListener(new Button.OnClickListener(){
                                          //打开 Wi-Fi
             public void onClick(View v) {
                OpenWifi(); //打开 wifi
             }
        });
        btnClose.setOnClickListener(new Button.OnClickListener(){
                                          //关闭 Wi-Fi
             public void onClick(View v) {
                 CloseWifi();
             }
        });
        btnSearch.setOnClickListener(new Button.OnClickListener(){  //搜索
             public void onClick(View v) {
                wifimanager.startScan();       //开始搜索
                wifilist=wifimanager.getScanResults();
                if(wifilist==null){
                    Toast.makeText(CH03_1.this,"没有无线网络可用", Toast.
LENGTH_LONG).show();
                }else{
                   ArrayAdapter<ScanResult>adapter=new ArrayAdapter <ScanResult>
( CH03_1.this, android.R.layout.simple_expandable_list_item_1, wifilist);
                   listview.setAdapter(adapter);
                }
            }
        });
    }
     public void OpenWifi(){
       if(!wifimanager.isWifiEnabled()){
          wifimanager.setWifiEnabled(true);
       }
    }
    public void CloseWifi(){
       if(wifimanager.isWifiEnabled()){
          wifimanager.setWifiEnabled(false);
       }
    }
}
```

3.1.6 蓝牙

蓝牙（Bluetooth）技术，实际上是一种短距离无线电技术，利用“蓝牙”技术，能够有

效地简化掌上电脑、笔记本式计算机和移动电话手持设备等移动通信终端设备之间的通信，也能够成功地简化以上这些设备与因特网 Internet 之间的通信，从而使这些现代通信设备与因特网之间的数据传输变得更加迅速高效，为无线通信拓宽道路。

蓝牙是无线数据和语音传输的开放式标准，它将各种通信设备、计算机及其终端设备、各种数字数据系统甚至家用电器采用无线方式连接起来。它的传输距离为 10 cm～10 m，如果增加功率或是加上某些外设便可达到 100 m 的传输距离。它采用 2.4 GHz ISM 频段和调频、跳频技术，使用权向纠错编码、ARQ、TDD 和基带协议。TDMA 每时隙为 0.625 μs，基带符合速率为 1Mbit/s。蓝牙支持 64 kbit/s 实时语音传输和数据传输，语音编码为 CVSD，发射功率分别为 1 mW、2.5 mW 和 100 mW，并使用全球统一的 48 位设备识别码。由于蓝牙采用无线接口代替有线电缆连接，具有很强的移植性，并且适用于多种场合，加上该技术功耗低、对人体危害小，而且应用简单、容易实现，所以易于推广。

蓝牙技术的系统结构分为三大部分：底层硬件模块、中间协议层和高层应用。

底层硬件部分包括无线跳频（RF）、基带（BB）和链路管理（LM）。无线跳频层通过 2.4 GHz 无须授权的 ISM 频段的微波，实现数据位流的过滤和传输，本层协议主要定义了蓝牙收发器在此频带正常工作所需要满足的条件。基带负责跳频以及蓝牙数据和信息帧的传输。链路管理负责连接、建立和拆除链路并进行安全控制。

蓝牙技术结合了电路交换与分组交换的特点，可以进行异步数据通信，可以支持多达 3 个同时进行的同步语音信道，还可以使用一个信道同时传送异步数据和同步语音。每个语音信道支持 64 kbit/s 的同步语音链路。异步信道可以支持一端最大速率为 721 kbit/s、另一端速率为 57.6 kbit/s 的不对称连接，也可以支持 43.2 kbit/s 的对称连接。

中间协议层包括逻辑链路控制和适应协议、服务发现协议、串口仿真协议和电话通信协议。逻辑链路控制和适应协议具有完成数据拆装、控制服务质量和复用协议的功能，该层协议是其他各层协议实现的基础。服务发现协议层为上层应用程序提供一种机制来发现网络中可用的服务及其特性。串口仿真协议层具有仿真 9 针 RS232 串口的功能。电话通信协议层则提供蓝牙设备间语音和数据的呼叫控制指令。

主机控制接口层（HCI）是蓝牙协议中软硬件之间的接口，它提供了一个调用基带、链路管理、状态和控制寄存器等硬件的统一命令接口。蓝牙设备之间进行通信时，HCI 以上的协议软件实体在主机上运行，而 HCI 以下的功能由蓝牙设备完成，二者之间通过一个对两端透明的传输层进行交互。

在蓝牙协议栈的最上部是各种高层应用框架。其中较典型的有拨号网络、耳机、局域网访问、文件传输等，它们分别对应一种应用模式。各种应用程序可以通过各自对应的应用模式实现无线通信。拨号网络应用可通过仿真串口访问微微网（Piconet），数据设备也可由此接入传统的局域网；用户可以通过协议栈中的 Audio（音频）层在手持设备和耳塞中实现音频流的无线传输；多台 PC 或笔记本式计算机之间不需要任何连线，就能快速、灵活地进行文件传输和共享信息，多台设备也可由此实现同步操作。

在 Android 系统中，提供了 android.bluetooth 包，里面提供了开发蓝牙应用所需要的主要接口。在表 3-1 中列出了几个常用的包：

表 3-1 蓝牙功能包

功 能 包	功 能
BluetoothAdapter	本地蓝牙设备的适配类，所有的蓝牙操作都需要通过该类完成
BluetoothDevice	蓝牙设备类，代表了蓝牙通讯过程中的远端设备
BluetoothSocket	蓝牙通信套接字，代表了与远端设备的连接点。使用 Socket 本地程序可以通过 inputstream 和 outputstream 与远端程序通信
BluetoothServerSocket	服务器通信套接字，与 TCP ServerSocket 类似
BluetoothClass	用于描述远端设备的类型、特点等信息，通过 getBluethoothClass()方法获取代表远端设备属性的 BluetoothClass 对象

就像前面讲过的 Wi-Fi 一样，在程序中需要获取相应的权限才能够对蓝牙进行操作。对蓝牙的操作权限如下：

```
<uses-permission android:name="android.permission.BLUETOOTH"/>
<user-permission android:name="android.permission.BLUETOOTH_ADMIN"/>
<uses-permission android:name="android.permission.READ_CONTACTS"/>
```

在设置足够的权限后对蓝牙进行操作，首先就是要取得蓝牙适配器，通过 BluetoothAdapter.getDefaultAdapter()方法获取本地的蓝牙适配器，正如表 3-2 中所描述的，若要获得远端的蓝牙适配器就要使用 BluetoothDevice 进行。在获取了本地蓝牙适配器后开启蓝牙，开启方法是通过一个 Intent 来进行，代码如下：

```
Intent startBT=new Intent(BluetoothAdapter.ACTION_REQUEST_ENABLE);
startActivityForResult(startBT,REQUEST_ENABLE);
```

同样的道理，通过 BluetoothAdapter 来请求系统允许蓝牙设备被搜索，只是 Action 的类型不一样而已：

```
Intent startSerach=new Intent(BluetoothAdapter.ACTION_REQUEST_DISCOVERABLE);
startActivityForResult(startSearch, REQUEST_DISCOVERABLE);
```

在表 3-2 中列出了常用的常量，用来执行某个活动、服务或者动作。

表 3-2 BluetoothAdapter 中动作常量

动 作 常 量	说 明
ACTION_DISCOVERY_FINISHED	完成蓝牙搜索
ACTION_DISCOVERY_STARTED	开始搜索
ACTION_LOCAL_NAME_CHANGED	更改蓝牙的名字
ACTION_REQUEST_DISCOVERABLE	请求能够被搜索
ACTION_REQUEST_ENABLE	请求启动蓝牙
ACTION_SCAN_MODE_CHANGED	扫描模式已改变
ACTION_STATE_CHANGED	状态已改变

除了上面的设置不要忘了要打开手持设备的蓝牙开关，当然在程序中通过调用 BluetoothAdapter 的 enable()方法来打开，通过调用 disable()方法来关闭。

下面介绍 BluetoothAdapter 中常用的一些方法，如表 3-3 所示。

表 3-3　BluetoothAdapter 中常用方法

常 用 方 法	返 回 值
public boolean cancelDiscovery ()	取消当前设备的发现查找进程
public static boolean checkBluetoothAddress (String address)	验证皆如"00:43:A8:23:10:F0"之类的蓝牙地址。字母必须为大写才有效。参数 address 为字符串形式的蓝牙模块地址返回值地址。正确则返回 true，否则返回 false
public boolean disable ()	关闭本地蓝牙适配器——不能在没有明确关闭蓝牙的用户动作中使用
public boolean enable ()	打开本地蓝牙适配器——不能在没有明确打开蓝牙的用户动作中使用
public String getAddress ()	返回本地蓝牙适配器的硬件地址。例如"00:11:22:AA:BB:CC"
public Set<BluetoothDevice> getBondedDevices ()	返回已经匹配到本地适配器的 BluetoothDevice 类的对象集合
public static synchronized BluetoothAdapter get DefaultAdapter ()	获取对默认本地蓝牙适配器的操作权限
public String getName ()	获取本地蓝牙适配器的蓝牙名称，这个名称对于外界蓝牙设备而言是可见的
public BluetoothDevice getRemoteDevice (String address)	为给予的蓝牙硬件地址获取一个 BluetoothDevice 对象
public int getScanMode()	获取本地蓝牙适配器的当前蓝牙扫描模式
public int getScanMode()	获取本地蓝牙适配器的当前蓝牙扫描模式
public int getState()	获取本地蓝牙适配器的当前状态
public boolean isDiscovering ()	如果当前蓝牙适配器正处于设备发现查找进程中，则返回真值
public boolean isEnabled ()	如果蓝牙正处于打开状态并可用，则返回真值与 getBluetooth State()==STATE_ON 等价
public BluetoothServerSocket listenUsingRfcomm WithServiceRecord(String name, UUID uuid)	创建一个正在监听的安全的带有服务记录的无线射频通信（RFCOMM）蓝牙端口
public boolean setName (String name)	设置蓝牙或者本地蓝牙适配器的昵称
public boolean startDiscovery ()	开始对远程设备进行查找的进程

蓝牙典型应用案例：搜索周围的蓝牙设备

该案例演示蓝牙的基本应用，搜索周围的蓝牙设备。由于模拟器不支持蓝牙模块,所以程序需要在真机上进行演示，程序运行后的主界面如图 3-2 所示。

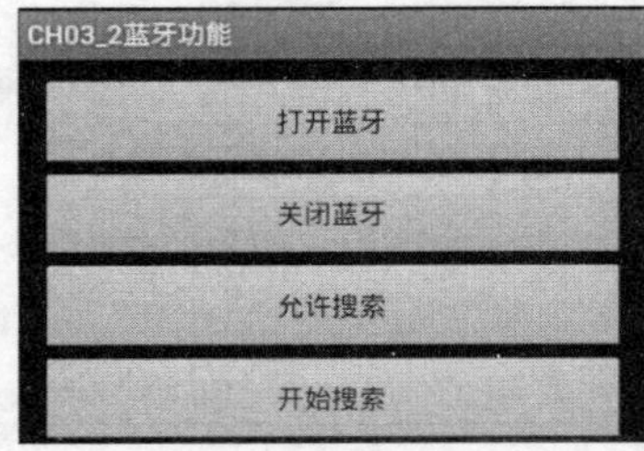

图 3-2　程序主界面

在主程序中没有复杂的知识，在打开和关闭蓝牙按钮的处理上通过前面介绍的知识进行处理。这里主要介绍“开始搜索”按钮所调用的程序。蓝牙设备搜索类的代码清单如下所示：CH03\CH03_2\src\sziit\ex\ch03\DiscoveryBT.java。

```
public class DiscoveryBT  extends ListActivity
{
   private Handler myHandler=new Handler();
   private BluetoothAdapter mBluetooth=BluetoothAdapter.getDefaultAdapter();
   //用来存储搜索到的蓝牙设备
   private List<BluetoothDevice> devices=new ArrayList<BluetoothDevice>();
```

```
    private volatile boolean isdiscoveryFinished;
    private Runnable isdiscoveryWorkder=new Runnable() {
      public void run()
      {  /* 开始搜索 */
        mBluetooth.startDiscovery();
        for (;;)
        {
          if (isdiscoveryFinished)
          {
            break;
          }
          try
          {
            Thread.sleep(100);
          }
          catch (InterruptedException e){}
        }
      }
    };
    /*** 接收器，当搜索蓝牙设备完成时调用*/
    private BroadcastReceiver mfoundReceiver=new BroadcastReceiver() {
      public void onReceive(Context context, Intent intent) {
        /* 从 intent 中取得搜索结果数据 */
        BluetoothDevice device=intent.getParcelableExtra(BluetoothDevice.
EXTRA_DEVICE);
        devices.add(device);          /* 将结果添加到列表中 */
        showDevices();                /* 显示列表 */
      }
    };
    private BroadcastReceiver isdiscoveryReceiver = new BroadcastReceiver()
    {
      public void onReceive(Context context, Intent intent)
      {  /* 卸载注册的接收器 */
        unregisterReceiver(mfoundReceiver);
        unregisterReceiver(this);
        isdiscoveryFinished=true;
      }
    };
    protected void onCreate(Bundle savedInstanceState)
    {
      super.onCreate(savedInstanceState);
      getWindow().setFlags(WindowManager.LayoutParams.FLAG_BLUR_BEHIND,
WindowManager.LayoutParams.FLAG_BLUR_BEHIND);
      setContentView(R.layout.discovery);
      if(!mBluetooth.isEnabled())     /* 如果蓝牙适配器没有打开，则结果 */
      {
        finish();
        return;
      }
      /* 注册接收器 */
```

```
        IntentFilter discoveryFilter=new IntentFilter(BluetoothAdapter.
ACTION_DISCOVERY_FINISHED);
        registerReceiver(isdiscoveryReceiver, discoveryFilter);
        IntentFilter foundFilter=new IntentFilter(BluetoothDevice.ACTION_FOUND);
        registerReceiver(mfoundReceiver, foundFilter);
        /* 显示一个对话框,正在搜索蓝牙设备 */
       SamplesUtils.indeterminate(DiscoveryBT.this, myHandler, "Scanning...",
isdiscoveryWorkder, new OnDismissListener() {
            public void onDismiss(DialogInterface dialog)
            {
                for (; mBluetooth.isDiscovering();)
                {
                    mBluetooth.cancelDiscovery();
                }
                isdiscoveryFinished=true;
            }
        }, true);
    }
    protected void showDevices()
    {   /* 显示设备列表 */
        List<String> list=new ArrayList<String>();
        for (int i=0, size=devices.size(); i<size; ++i)
        {
            StringBuilder b=new StringBuilder();
            BluetoothDevice d=devices.get(i);
            b.append(d.getAddress());
            b.append('\n');
            b.append(d.getName());
            String s=b.toString();
            list.add(s);
        }
        final ArrayAdapter<String> adapter=new ArrayAdapter<String>(this,
android.R.layout.simple_list_item_1, list);
        myHandler.post(new Runnable() {
            public void run()
            {
                setListAdapter(adapter);
            }
        });
    }
    protected void onListItemClick(ListView l,View v,int position,long id)
    {
        Intent result=new Intent();
        result.putExtra(BluetoothDevice.EXTRA_DEVICE, devices.get(position));
        setResult(RESULT_OK, result);
        finish();
    }
}
```

首先用 getDefaultAdapter()方法取得默认的蓝牙适配器，并且创建了一个用来存储搜索到蓝牙设备的 List，然后在程序开始时注册了搜索已完成和发现设备两个 BroadcastReceiver。然后通过一个线程来控制蓝牙设备的搜索，当搜索中有触发两个接收器的事件时，就直接传递给接收器进行保存。最后将保存在 List 中的蓝牙设备显示在 ListView 中。

3.2 技术准备

3.2.1 Android 网络基础

现实生活中有许多形形色色的“网络”，如看得见的有电网、公共电话网、水网、有线电视网等，看不见的有销售网、人际关系网等，众多的网络组成了人们的生存空间。目前最热门、发展速度最快的无疑是计算机网络。

计算机网络是现代通信技术与计算机技术相结合的产物，它利用网卡、网线、交换机等网络设备，把分散在各地的计算机连接起来，并通过特定的软件（网络协议）实现计算机之间的相互通信和资源共享。

从不同角度出发，计算机网络有多种分类方法，常见的分类有：

- 按覆盖范围分类：局域网（Local Area Network，LAN）、城域网（Metropolis Area Network，MAN）和广域网（Wide Area Network，WAN）。
- 按网络拓扑结构分类：星状网络、环状网络、总线状网络等。

3 种网络接口为：标准 Java 接口、Apache 接口、Android 网络接口。

Android 目前有 3 种网络接口可以使用，分别为：java.net.*、org.apache 和 android.net.*。下面简单介绍一下这些接口的功能和作用。

1. **标准 Java 接口**

Java.net.*提供与网络连接相关的类，java.net.*的包分为两部分：低级 API 和高级 API。

低级 API 主要用于处理以下抽象：

- 地址，也就是网络标识符，如 IP 地址。
- 套接字，也就是基本双向数据通信机制。
- 接口，用于描述网络接口。

高级 API 主要用于处理以下抽象：

- URI，表示统一资源标识符（Uniform Resource Identifier）。Web 上可用的每种资源（HTML 文档、图像、视频片段、程序等）可通过 URI 来定位。
- URL，表示统一资源定位符（Uniform Resource Locator），也被称为网页地址，是因特网上标准的资源地址。
- 连接，表示到 URL 所指向资源的连接。
- 地址，在整个 java.net API 中，地址或者用作主机标识符或者用作套接字端点标识符。

下面通过一段程序代码说明 java.net.*在程序中的使用。

使用 java.net 创建连接：

```
try
{
   URL url=new URL("http://www.baidu.com");    //定义 URL 标识符
   HttpURLConnection http=(HttpURLConnection) url.openConnection();
                                               //打开连接
```

```
    int nRC=http.getResponseCode();                    //得到连接状态
    if(nRC==HttpURLConnection.HTTP_OK){
       InputStream is=http.getInputStream();      /取得数据
       ……//处理数据
    }
}catch(Exception e){
}
```

2. Apache 接口

HTTP 是目前在 Internet 上使用最多、最重要的通信协议，越来越多的 Java 应用程序需要通过 HTTP 访问网络资源。虽然前面已经讲了 java.net 包中已经提供了访问 HTTP 的基本功能，但是这对于大部分应用程序是不够的。Android 系统引入了 Apache HttpClient 以及对其的封装和扩展，如设置缺省的 HTTP 超时和缓存大小等。Android 使用的是目前最新的 HttpClient 4.0。通过 Apache 创建 HttpClient 以及 Get/Post、HttpRequest 等对象，设置连接参数，执行 HTTP 操作，处理服务器返回结果等功能。下面同样通过代码段说明 Apache 接口的应用。

使用 android.net.http.*连接网络：

```
try{
   HttpClient hc=new DefaultHttpClient();        //创建HttpClient使用默认属性
   HttpGet get=new HttpGet("http://www.baidu.com"); //创建HttpGet实例
   HttpResponse rp=hc.execute(get);                //连接
   if(rp.getStatusLine().getStatusCode()==HttpStatus.SC_OK){
      InputStream is=rp.getEntity().getContent();
      ……//处理数据
   }
}catch(IOException )
}
```

3. Android 网络接口

Android.net.*包实际上是通过对 Apache 的 HttpClient 进行封装，实现的一个 HTTP 变成接口，同时也提供了 HTTP 请求队列管理以及 HTTP 连接池管理，以提高并发情况下的处理效率，除此之外还有网络状态监视等接口、网络访问的 Socket，常用的 Uri 类以及有关 WI-FI 相关的类等。

使用 Android 中的 Socket 连接网络：

```
try{
     InetAddress inetAddress=InetAddress.getByName("192.168.1.25");
     Socket client=new Socket(inetAddress,61203,true);
     InputStream in=client.getInputStream();
     OutputStream out=client.getOutputStream();      //处理数据
     out.close();
     in.close();
     client.close();
}catch(UnknownHostException e){
}catch(IOException e){
}
```

3.2.2 HTTP 通信

1. HTTP Get 与 HTTP Post

HTTP 通信中使用最多的就是 Get 和 Post。Get 请求方式中，参数直接放在 URL 字串后面，传递给服务器。格式如下：

```
HttpGet method=new HttpGet("http://www.baidu.com?admin=Get");
HttpResponse response=client.execute(method);
```

而 Post 请求方式中，参数必须采用 NameValuePair[]数组的传送方式。格式如下：

```
HttpPost method=new HttpPost("http://www.baidu.com");
List<NameValuePair> params=new ArrayList<NameValuePair>();
params.add(new BasicNameValuePair("admin","Get"));
method.setEntity(new UrlEncodedFormEntity(params));
HttpResponse response=client.execute(method);
```

在这两种通信方式中，一般情况下，两种方式实现的效果一样。但也有特殊情况，可能服务器只支持 Get 请求方式，而不支持 Post 请求方式，所以导致 Post 请求方式获取不到需要的数据；也可能服务器只支持 Post 请求方式，不支持 Get 请求方式。于是，需要查看服务器返回的状态码，如果是 200 则证明连接成功，否则连接失败。状态码的取得方式可以通过抓包观察，也可以直接用代码获取。

用代码获得服务器返回的状态码，具体代码如下：

```
HttpResponse httpResponse=new DefaultHttpClient().execute(method);
If(httpResponse.getStatusLine().getStatusCode()==200)
{ //从 URL 获取数据
}
else
{ //显示连接错误
}
```

在这里需要注意的是，由于 Android 的很多操作都涉及权限问题，比如打电话和发短信等，都需要权限。而 Android 尝试连续网络时，也需要权限。在 AndroidManifest.xml 中添加加入网络连接权限：

```
<uses-permission android:name="android.permission.INTERNET" />
```

2. HttpURLConnection 接口

在 Android 的 SDK 中，Google 同时也继承了网络连接中标准的 Java 接口，使得最基本的一些连接方式得以继续沿用。注意，URLConnection 与 HttpURLConnection 都是抽象类，无法直接实例化对象。其对象主要通过 URL 的 openConnection()方法获得。

标准的 Java 接口格式如下：

```
URL url=new URL("http://www.baidu.com");
HttpURLConnection http=(HttpURLConnection)url.openConnection();
int response=http.getResponseCode();
if(200==response)
{//从 URL 获取数据
}
else
{//显示连接错误
}
```

在上述这些方法中，如果所连接的网址不存在，会报出 java.net.UnknownHostException 异常，所以连接需要 try catch。

3. 网络接口

Android 中的网络接口实际上是通过对 Apache 中 HttpClient 的封装来实现的一个 HTTP 编程接口。例如：

```
InetAddress ia=InetAddress.getByName("192.168.1.100");
Socket client=new Socket(ia, 8082, true);
```

该例子已不适用于 Android，因为在 SDK 中，Socket 的构造方法 Socket(InetAddress addr, int port, boolean streaming) 被定义为：This constructor is deprecated。在 Android 中，被弃用的方法都调试不通，会报异常，所以该构造方法不能使用。

此处说明的是，因为在以前 Java 代码中（非 Android），有部分被定义为 This constructor is deprecated 的方法，被调用时并不会报异常，而使得部分程序员继续使用，在 Android 中是行不通的。其实采用最简单的 Socket 构造方法即可，例如 Socket clientSocket=new Socket ("10.12.39.25",9000);

4. 权限验证

首先需要明确的是，权限验证不同于参数传递。例如，在访问网页的时候，如果需要翻页，有的会把页数 id 这个参数传递过去，通过这个 id 指向对应的页面。而权限验证比如是在路由器下，想访问 192.168.0.1 时，会要求输入账号和密码，这就不能通过参数传递的方法。权限验证的方法如下：

```
DefaultHttpClient client=new DefaultHttpClient();
client.getCredentialsProvider().setCredentials(AuthScope.ANY,
new UsernamePasswordCredentials("admin", "password"));
```

通过上述方法，把账号和密码传递给服务器进行验证，即可成功连接。

下面通过一个简单的实例来了解 HTTP 通信中 HTTP Get 与 HTTP Post 的区别。这个实例中有不同的获取网页上数据的方式。在这个例子中需要建立两个 Web 页面作为支持，并需要在本地配置服务器，这里使用 Tomcat 作为服务器。对于 Web 页面的建立以及其运行方法这里不再赘述，读者可以自己了解。两个页面在浏览器中的访问效果如图 3-3 和图 3-4 所示。

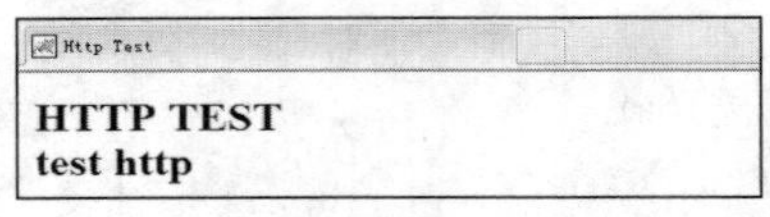

图 3-3 http1.jsp 效果

My JSP 'httpget.jsp' starting page

our parameters:OUR INPUT

图 3-4 httpget.jsp 效果

在完成了 Web 页面后就来看一下在手持设备上对网页的访问效果，分别试验三种不同的访问方式，第一种是直接访问，第二种是通过 Get 方式访问 httpget.jsp，第三种是通过 Post 方式访问 httpget.jsp 页面。主程序的布局如图 3-5 所示。

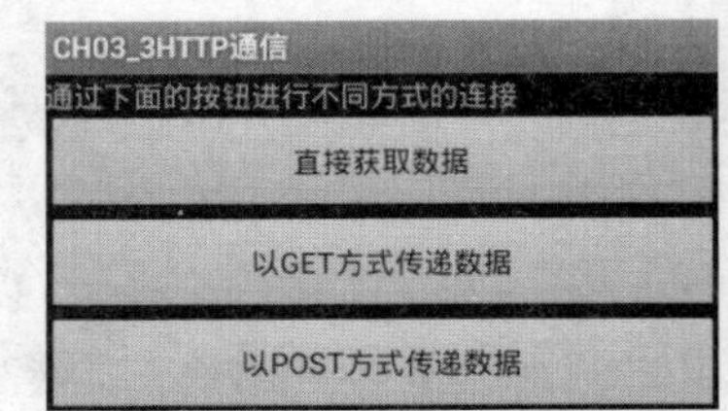

图 3-5 主程序布局效果

主程序布局文件代码如下：CH03\CH03_03\res\layout\main.xml。

```
<?xml version="1.0" encoding="utf-8"?>
<LinearLayout xmlns:android="http://schemas.android.com/apk/res/android"
```

```
    android:orientation="vertical"
    android:layout_width="fill_parent"
    android:layout_height="fill_parent">
<TextView
      android:layout_width="fill_parent"
      android:layout_height="wrap_content"
      android:text="通过下面的按钮进行不同方式的连接"/>
   <Button
      android:id="@+id/Button_HTTP"
      android:layout_width="fill_parent"
      android:layout_height="wrap_content"
      android:text="直接获取数据"/>
   <Button
      android:id="@+id/Button_Get"
      android:layout_width="fill_parent"
      android:layout_height="wrap_content"
      android:text="以 GET 方式传递数据"/>
   <Button
      android:id="@+id/Button_Post"
      android:layout_width="fill_parent"
      android:layout_height="wrap_content"
      android:text="以 POST 方式传递数据"/>
</LinearLayout>
```

然后来看看访问不需要传递参数的网页如何实现，只需要打开一个 HttpUrlConnection 连接，然后取得流中的数据，完成后关闭这个连接，实现代码如下：CH03\CH03_03\src\sziit\ex\ch03\Activity02.java。

```
public class Activity02 extends Activity
{
   private final String DEBUG_TAG="Activity02";
   public void onCreate(Bundle savedInstanceState)
                                      /** 第一次创建活动（activity ）时调用*/
   {
      super.onCreate(savedInstanceState);
      setContentView(R.layout.http);
      TextView mTextView=(TextView)this.findViewById(R.id.TextView_ HTTP);
      String httpUrl="http://192.168.2.102:8089/Android/http1.jsp";
                                      //http 地址
      String resultData="";           //获得的数据
      URL url=null;
      try
      {  //构造一个 URL 对象
         url=new URL(httpUrl);
      }
      catch(MalformedURLException e)
      {
         Log.e(DEBUG_TAG,"MalformedURLException");
      }
```

```
        if (url != null)
        {
           try
           {  //使用 HttpURLConnection 打开连接
              HttpURLConnection urlConn=(HttpURLConnection) url.openConnection();
              //得到读取的内容(流)
              InputStreamReader in=new InputStreamReader(urlConn.getInputStream());
              // 为输出创建 BufferedReader
              BufferedReader buffer=new BufferedReader(in);
              String inputLine=null;
              while(((inputLine=buffer.readLine()) != null))
                                          //使用循环来读取获得的数据
              {  //在每一行后面加上一个"\n"来换行
                 resultData+=inputLine+"\n";
              }
              in.close();                 //关闭 InputStreamReader
              urlConn.disconnect();       //关闭 http 连接
              if ( resultData != null )
              {  //设置显示取得的内容
                 mTextView.setText(resultData);
              }
              else
              {
                 mTextView.setText("读取的内容为 NULL");
              }
           }
           catch (IOException e)
           {
              Log.e(DEBUG_TAG, "IOException");
           }
        }
        else
        {
           Log.e(DEBUG_TAG, "Url NULL");
        }
     }
  //设置返回按键监听，在此省略
  }
```

在定义 HttpUrl 的时候需要特别注意，在浏览器中访问本地页面时可以用 localhost 替代本机的 IP 地址，但是在 Android 模拟器中如果使用的是 localhost，则此时代表的是模拟器本身，所以这里需要用具体机器的 IP 地址。可以通过命令 ipconfig/all 查看本机的 IP 地址，方法如图 3-6 所示。

运行程序，效果如图 3-7 所示。

对于通过 Get 方式访问 httpget.jsp，只需要在 Activity02 的基础上进行修改即可：即将 httpUrl 改成：String httpUrl = http://192.168.2.102:8089/Android/httpget.jsp?par=1234567;，需要注意的是，这里的“?par=1234567”是需要传递的参数 par。修改后运行效果如图 3-8 所示。

```
C:\Documents and Settings\Administrator>ipconfig/all

Windows IP Configuration

        Host Name . . . . . . . . . . . . : terry
        Primary Dns Suffix  . . . . . . . :
        Node Type . . . . . . . . . . . . : Mixed
        IP Routing Enabled. . . . . . . . : No
        WINS Proxy Enabled. . . . . . . . : No

Ethernet adapter 本地连接:

        Media State . . . . . . . . . . . : Media disconnected
        Description . . . . . . . . . . . : Realtek PCIe GBE Family Controller
        Physical Address. . . . . . . . . : 00-26-9E-92-E3-06

Ethernet adapter 无线网络连接:

        Connection-specific DNS Suffix  . :
        Description . . . . . . . . . . . : 11b/g/n  Wireless LAN Mini-PCI Expr
        Physical Address. . . . . . . . . : 00-26-5E-ED-66-0C
        Dhcp Enabled. . . . . . . . . . . : Yes
        Autoconfiguration Enabled . . . . : Yes
        IP Address. . . . . . . . . . . . : 192.168.2.102
        Subnet Mask . . . . . . . . . . . : 255.255.255.0
```

图 3-6　通过 ipconfig/all 命令获得本机 IP 地址

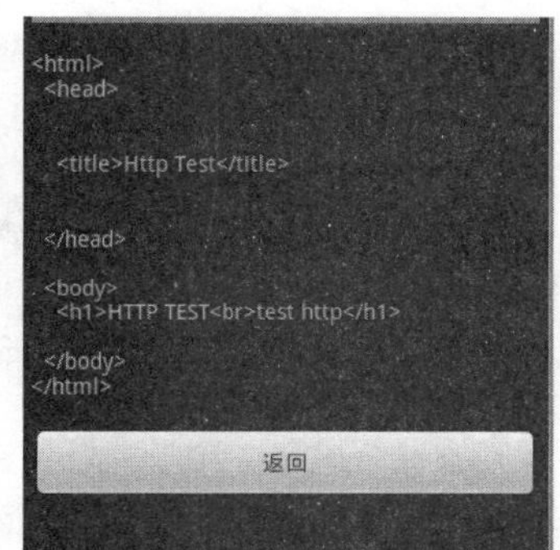

图 3-7　直接访问 http1.jsp

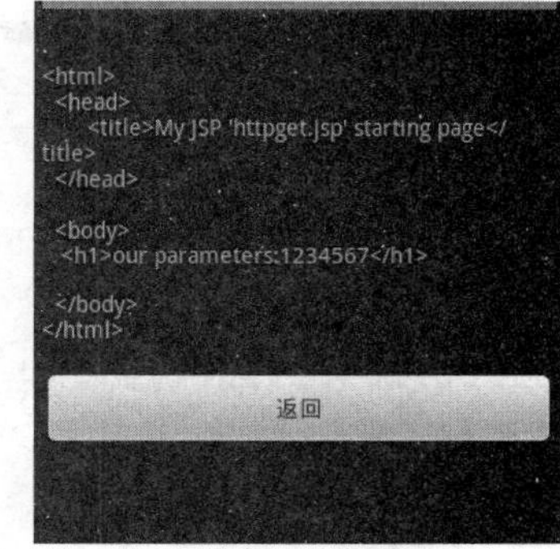

图 3-8　Get 方式访问 httpget.jsp

最后来了解 Post 方式，因为 HttpURLConnection 默认使用 Get 方式，所以在使用 Post 方式的时候需要对请求的方式进行设置，即 setRequestMethod("POST")方法进行设置。然后将要传递的参数内容通过 writeBytes()方法写入数据流。具体的实现代码如下：CH03\CH03_03\src\sziit\ex\ch03\Activity04.java。

```
public class Activity04 extends Activity {
  private final String DEBUG_TAG="Activity04";
  public void onCreate(Bundle savedInstanceState)
  /** 第一次创建活动（activity ）时调用*/
  {
     super.onCreate(savedInstanceState);
     setContentView(R.layout.http);
     TextView mTextView=(TextView)this.findViewById(R.id.TextView_HTTP);
     //http 地址“?par=abcdefg”是上传的参数
     String httpUrl="http://192.168.2.102:8089/Android/httpget.jsp";
     String resultData="";          //获得的数据
     URL url=null;
     try
     {  //构造一个 URL 对象
        url=new URL(httpUrl);
     }
     catch(MalformedURLException e)
```

```
        {
            Log.e(DEBUG_TAG, "MalformedURLException");
        }
        if(url != null)
        {
            try
            {
                //使用 HttpURLConnection 打开连接
                HttpURLConnection urlConn = (HttpURLConnection) url.openConnection();
                //因为这个是 Post 请求，设立需要设置为 true
                urlConn.setDoOutput(true);
                urlConn.setDoInput(true);
                urlConn.setRequestMethod("POST"); // 设置以 Post 方式
                urlConn.setUseCaches(false);       // Post 请求不能使用缓存
                urlConn.setInstanceFollowRedirects(true);
                //配置本次连接的 Content-type，配置为 application/x-www-form-
urlencoded 的
                urlConn.setRequestProperty("Content-Type","application/x-www-
form-urlencoded");
                //连接，从 postUrl.openConnection()至此的配置必须要在 connect 之前
完成，要注意的是 connection.getOutputStream 会隐含的进行 connect
                urlConn.connect();
                //DataOutputStream 流
                DataOutputStream out=new DataOutputStream(urlConn.getOut
putStream());
                 //要上传的参数
                 String content = "par=" + URLEncoder.encode("ABCDEFG", "gb2312");
                 out.writeBytes(content); //将要上传的内容写入流中
                 out.flush();                //刷新、关闭
                 out.close();
                 //获取数据
                 BufferedReader reader=new BufferedReader(new InputStreamReader
(urlConn.getInputStream()));
                String inputLine=null;
                //使用循环来读取获得的数据
                while(((inputLine=reader.readLine()) != null))
                { //在每一行后面加上一个"\n"来换行
                    resultData+=inputLine+"\n";
                }
                reader.close();
                urlConn.disconnect();//关闭 http 连接
                if(resultData != null )
                {  //设置显示取得的内容
                    mTextView.setText(resultData);
                }
                else
                {
                    mTextView.setText("读取的内容为 NULL");
                }
            }
```

```
            catch (IOException e)
            {
               Log.e(DEBUG_TAG, "IOException");
            }
         }
         else
         {
            Log.e(DEBUG_TAG, "Url NULL");
         }
         //设置返回按键监听，在此省略
      }
   }
```

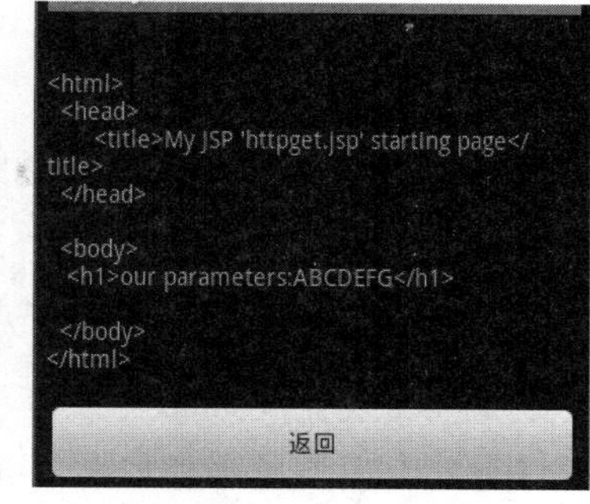

图 3-9　Post 方式访问 httpget.jsp

程序运行效果如图 3-9 所示。

3.2.3　Socket 通信

所谓 Socket 通常也称作“套接字”，应用程序通常通过“套接字”向网络发出请求或者应答网络请求，用于描述 IP 地址和端口，是一个通信链的句柄。

常用的 socket 有两种类型，分别是：流式 Socket（SOCK_STREAM）和数据报式 Socket（SOCK_DGRAM）。流式是一种面向连接的 Socket，针对于面向连接的 TCP（Transmission Control Protocol，传输控制协议）服务应用；数据报式 Socket 是一种无连接的 Socket，对应于无连接的 UDP（User Datagram Protocol，用户数据报协议）服务应用。

Socket 通信的原理比较简单，大致分为以下几个步骤：

服务器端步骤：

（1）建立服务器端的 Socket，开始侦听整个网络的连接请求。

（2）当检测到来自客户端的请求时，向客户端发送收到连接请求的信息，并与之建立连接。

（3）当通信完成后，服务器关闭与客户端的 Socket 连接。

客户端步骤：

（1）建立客户端的 Socket，确定要连接的服务器的主机名和端口。

（2）发送连接请求到服务器，并等待服务器的回馈信息。

（3）建立连接后，与服务器开始进行数据传输与交互。

（4）数据处理完毕后，关闭自身的 Socket 连接。

在 Java 中，在 java.net 包中提供了两个类 Socket 和 ServerSocket。ServerSocket 用于服务器端，Socket 是建立网络连接时使用的。在连接成功时，应用程序两端都会产生一个 Socket 实例，操作这个实例，完成所需的会话。对于一个网络连接来说，套接字是平等的，并没有差别，不因为在服务器端或在客户端而产生不同级别。不管是 Socket 还是 ServerSocket 它们的工作都是通过 SocketImpl 类及其子类完成的。下面了解一下 Socket 的构造方法：

- Socket(InetAddress address,int port);
- Socket(InetAddress address,int port,boolean stream);
- Socket(String host,int port);
- Socket(String host,int port,boolean stream);
- Socket(SocketImpl impl);

- Socket(String host,int port,InetAddress localAddr,int localPort);
- Socket(InetAddress address,int port,InetAddress localAddr, int localport);
- ServerSocket(int port);
- ServerSocket(int port,int backlog);
- ServerSocket(int port,int backlog,InetAddress bindAddr);

其中，address、host 和 post 分别是双向连接中另一方的 IP 地址、主机号和端口号，stream 指明 Socket 是流式 Socket 还是数据报式 Socket，localPort 表示本地主机的端口号，loalAddr 和 bindAddr 是本机器的地址，impl 是 Socke 的父类，既可以用来创建 ServerSocket，又可以用来创建 Socket。

下面介绍几个常用的方法：

- accept()方法用于产生“阻塞”，直到接受到一个连接，并且返回一个客户端的 Socket 对象实例。“阻塞”是一个术语，它使程序运行暂时“停留”在这个地方，直到一个会话产生，然后程序继续；通常“阻塞”是由循环产生的。
- getInputStream()方法获得网络连接输入，同时返回一个 InputStream 对象实例。
- getOutputStream()方法连接的另一端将得到输入，同时返回一个 OutputStream 对象实例。

注意：其中 getInputStream 和 getOutputStream 方法均会产生一个 IOException，它必须被捕获，因为它们返回的流对象，通常都会被另一个流对象使用。

最后，在每一个 Socket 对象使用完毕时，要使其关闭，使用 Socket 对象的 close()方法，在关闭 Socket 之前，应将与 Socket 相关的所有输入流和输出流关闭，以释放所有资源。先关闭输入流和输出流，再关闭 Socket。

1. 简易通信

在 Android 中完全可使用 Java 标准 API 来开发应用，下面实现一个服务器与客户端的通信，服务器用 Java 工程实现，客户端用 Android 工程实现。服务器端代码如下：

```
   public class Server {
     public static void main(String[] args) throws IOException {
        //创建 ServerSocket 服务器对象，端口号设置为 0122
        ServerSocket server=new ServerSocket(0122);
        Socket client=server.accept();     //创建 Socket 对象，负责与客户通信
        BufferedReader in=new BufferedReader(new InputStreamReader(client
.getInputStream()));                       //创建输入流对象
        PrintWriter out=new PrintWriter(client.getOutputStream());
                                           //创建输出流对象
        while (true) {
           String str=in.readLine();
           if(str==null)
              break;
           System.out.println(str);
           out.println("Server receive"+str);
           out.flush();
           if(str.equals("over"))
              break;
        }
        out.close();
```

```
            in.close();
            client.close();
        }
    }
```

这里需要注意的是不要在关闭 Writer 之前关闭 Reader。

客户端实现源码代码如下：

```
public class ClientActivity extends Activity {
    private Button btn;
    private EditText sendData;
    private TextView getData;
    private Socket server;
     public void onCreate(Bundle savedInstanceState) {
        super.onCreate(savedInstanceState);
        setContentView(R.layout.main);
        btn=(Button)findViewById(R.id.btn);
        sendData=(EditText)findViewById(R.id.setData);
        getData=(TextView)findViewById(R.id.getData);
        btn.setOnClickListener(new Button.OnClickListener() {
         public void onClick(View v) {
            try {
               server=new Socket("192.168.1.47", 0122);
               BufferedReader in=new BufferedReader(new InputStreamReader
(server.getInputStream()));
               PrintWriter out=new PrintWriter(server.getOutputStream());
               String str=sendData.getText().toString();
                  out.println(str);
                  out.flush();
                  getData.setText(in.readLine());
                  out.close();
                  in.close();
                  server.close();
            } catch (UnknownHostException e) {
               e.printStackTrace();
            } catch (IOException e) {
               e.printStackTrace();
            }
         }
      });
    }
}
```

启动服务器之后，运行 Android Socket 的客户端代码，模拟器如图 3-10 所示。

当随便输入一个信息（如 123）后，单击“获取服务器数据”按钮，服务器会传回信息并显示出来，如图 3-11 所示。

在这个简易 Socket 通信中，当客户端发送信息给服务器，然后从服务器接收到返回信息后，客户端会切断和服务器的连接，即只能通信一次。

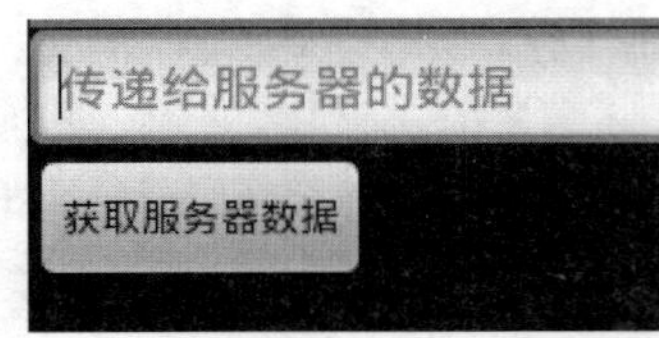

图 3-10　程序运行效果

图 3-11　获取服务器数据

2. Socket 高级通信

前面学习的 Socket 简易通信中，通信一次即断掉并不是想要的效果。需要修改服务器，使其与客户端通信是否断掉由客户端决定，而不是只能通信一次。并且高级通信可以使服务器能同时与多个客户端相互连接。下面对原代码进行修改，使其能达到想要的效果。在这个例子中可以在控制台中对输入的信息进行测试，也可以用上面实例中的客户端程序进行测试。

服务器实现代码如下：

```
public class Server {
    public static void main(String[] args) throws IOException {
        try {//创建 ServerSocket 服务器对象，端口号设置为 0122
            ServerSocket server=new ServerSocket(0122);
            while(true){
                Socket client; //创建 Socket 对象，负责与客户端通信
                client=server.accept();
                ReceServer rs=new ReceServer(client);
                rs.start();
                Thread.currentThread().sleep(500);
            }
        } catch (InterruptedException e) {
            e.printStackTrace();
        } catch (IOException e) {
            e.printStackTrace();
        }
    }
}
class ReceServer extends Thread{
    private Socket client;
    ReceServer(Socket client){
        this.client=client;
    }
    public void run() {
        try { //创建输入流对象
            BufferedReader in=new BufferedReader(new InputStreamReader
(client.getInputStream()));
            PrintWriter out=new PrintWriter(client.getOutputStream());
                //创建输出流对象
            while (true) {
                String str=in.readLine();
                if(str==null)
                    break;
                System.out.println(str);
                out.println("Server receive"+str);
                out.flush();
                if (str.equals("over"))
```

```
                break;
            }
            out.close();
            in.close();
            client.close();
        } catch (IOException e) {
            e.printStackTrace();
        }
    }
}
```

客户端实现的代码基本不用修改，直接采用简易通信中的客户端实现代码即可。在模拟器中运行了客户端应用之后，客户端可以与服务器进行多次交互，并不是一次交互之后就断掉，如图 3-12 和图 3-13 所示。

图 3-12　客户端传输一次信息

图 3-13　客户端继续传输信息

高级通信实现了多个客户端同时可以与服务器进行交互，可以开两个模拟器，同时运行客户端应用，发现两个客户端通信互不影响。

3.3 案例

3.3.1 迷你浏览器

该迷你浏览器实现嵌入式设备（如手机、移动设备和平板电脑）等上的浏览器功能，它包括两个组件，一个文本框用于接收用户输入想访问的 URL；一个 WebView 用于加载并显示该 URL 对应的页面。该程序的布局效果如图 3-14 所示，其详细代码请参见配套光盘中的 CH03_4。

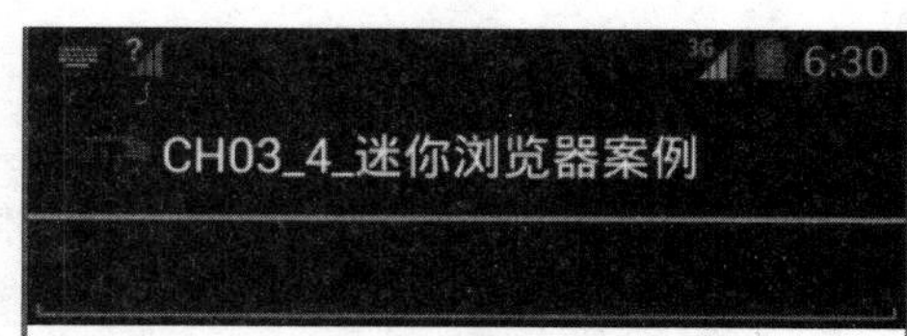

图 3-14　迷你浏览器案例布局主界面

程序的界面布局代码如下：CH03 \CH03_4\res\layout\ activity_main.xml。

```
    <?xml version="1.0" encoding="utf-8"?>
<LinearLayout xmlns:android="http://schemas.android.com/apk/res/android"
    android:orientation="vertical"
    android:layout_width="fill_parent"
    android:layout_height="fill_parent">
<EditText
    android:id="@+id/url"
    android:layout_width="fill_parent"
    android:layout_height="wrap_content" />
<!-- 显示页面的 WebView 组件 -->
<WebView
```

```
    android:id="@+id/show"
    android:layout_width="fill_parent"
    android:layout_height="fill_parent"/>
</LinearLayout>
```

程序的 Java 代码：CH03\CH03_4\src\com\zigcloud\browser\MiniBrowser.java。

```
public class MiniBrowser extends Activity
{
    EditText url;
    WebView show;
    public void onCreate(Bundle savedInstanceState)
    {
        super.onCreate(savedInstanceState);
        setContentView(R.layout.activity_main);
        //获取页面中文本框、WebView组件
        url=(EditText) findViewById(R.id.url);
        show=(WebView) findViewById(R.id.show);
    }
    public boolean onKeyDown(int keyCode, KeyEvent event)
    {
        if(keyCode==KeyEvent.KEYCODE_SEARCH)
        {
            String urlStr=url.getText().toString();
            show.loadUrl(urlStr);         // 加载、并显示urlStr对应的网页
            return true;
        }
        return false;
    }
}
```

该程序的运行效果如图 3-15 所示。

图 3-15 迷你浏览器案例运行主界面

3.3.2 获取 Web 网址信息

在开发一些基于 Web 系统的 Android 客户端程序开发过程中，通常要获取服务器数据是需要访问一个 Web 网址来获取的，再将内容进行解析、处理和显示的，本案例主要实现获取 Web 网址信息的功能，如从气象 Web 网址获取天气数据，从物联网实训台获取设备数据等，该程序的界面布局如图 3-16 所示，其详细代码请参见配套光盘中的 CH03_5。

图 3-16 获取 Web 网址信息案例布局主界面

程序的界面布局代码如下：CH03\CH03_5\res\ layout\activity_main.xml。

```
<RelativeLayout xmlns:android="http://schemas. android.com/apk/res/android"
    xmlns:tools="http://schemas.android.com/tools"
    android:layout_width="match_parent"
    android:layout_height="match_parent"
    android:paddingBottom="@dimen/activity_vertical_margin"
    android:paddingLeft="@dimen/activity_horizontal_margin"
    android:paddingRight="@dimen/activity_horizontal_margin"
    android:paddingTop="@dimen/activity_vertical_margin"
    tools:context=".MainActivity" >
```

```
    <Button
        android:id="@+id/btn_getdata_weather"
        android:layout_width="match_parent"
        android:layout_height="wrap_content"
        android:layout_alignParentLeft="true"
        android:layout_alignParentTop="true"
        android:text="@string/getdata_weather" />
    <Button
        android:id="@+id/btn_equipdatas"
        android:layout_width="match_parent"
        android:layout_height="wrap_content"
        android:layout_alignParentLeft="true"
        android:layout_below="@+id/btn_getdata_weather"
        android:text="@string/getdata_equipdatas" />
    <TextView
        android:id="@+id/tv_content"
        android:layout_width="wrap_content"
        android:layout_height="wrap_content"
        android:layout_alignLeft="@+id/btn_getdata_weather"
        android:layout_alignParentBottom="true"
        android:layout_alignParentRight="true"
        android:layout_below="@+id/btn_equipdatas"
        android:scrollHorizontally="true"
        android:text="" />
</RelativeLayout>
```

程序的 Java 代码如下：CH03\CH03_5\src\ com\example\ MainActivity.java。

```
public class MainActivity extends Activity {/***主页面* */
    protected void onCreate(Bundle savedInstanceState) {
        super.onCreate(savedInstanceState);
        setContentView(R.layout.activity_main);
        //设置获取天气数据的按钮监听器
        //地址: http://m.weather.com.cn/data/101250101.html
        findViewById(R.id.btn_getdata_weather).setOnClickListener(new View.
OnClickListener() {
            public void onClick(View v) {
                getWeather();
            }
        });
        //设置获取实训台设备的数据
        //地址: http://www.iotcase.com:8888//bin/node.cgi?method=get
        findViewById(R.id.btn_equipdatas).setOnClickListener(new View.
OnClickListener() {
            public void onClick(View v) {
                getEquipDatas();
            }
        });
    }
    public void getWeather(){
        new Thread(){
            public void run() {
                super.run();
                String res="";
                try {
```

```
                res=HttpUtil.getRequest("http://m.weather.com.cn/data/
101250101.html");
            } catch (Exception e) {
                res="出现异常: "+e.getMessage();
            }
            setText(res);
        }
    }.start();
}
public void getEquipDatas(){
    new Thread(){
        public void run() {
            super.run();
            String res="";
            try {
              //res=HttpUtil.getRequest("http://www.iotcase.com:8888//bin/node.
cgi?method=get");
                res=HttpUtil.getRequest("http://192.168.1.201//bin/
node.cgi?method=get");
            } catch (Exception e) {
                res="出现异常: "+e.getMessage();
            }
            Log.v("log", res);
            setText(res);
        }
    }.start();
}
private Handler mHandler=new Handler();
private void setText(final String str){ /*** 设置文本框内容* */
    final TextView tv_content=(TextView)findViewById(R.id.tv_content);
    mHandler.post(new Runnable() {
        public void run() {
            tv_content.setText(str);
        }
    });
}
public boolean onCreateOptionsMenu(Menu menu) {
    getMenuInflater().inflate(R.menu.main, menu);
                                    //膨胀的菜单，将菜单项添加到行动条
    return true;
}
}
```

程序清单 CH03\ CH03_5\ com\example\demohttpclient\HttpUtil.java。

```
public class HttpUtil /*** Http 封装类*/
{ /***创建 HttpClient 对象* */
   public static HttpClient httpClient=new DefaultHttpClient();
   /***
    * @param url 发送请求的 URL
    * @return 服务器响应字符串
    * @throws Exception
    */
   public static String getRequest(String url)
      throws Exception{
      HttpGet get=new HttpGet(url);          //创建 HttpGet 对象
```

```
        HttpResponse httpResponse=httpClient.execute(get);  //发送 GET 请求
        // 如果服务器成功地返回响应
        if(httpResponse.getStatusLine().getStatusCode()==200){
            //获取服务器响应字符串
            String result=EntityUtils   .toString(httpResponse.getEntity());
            return result;
        }
        return null;
    }
    /*** @param url 发送请求的 URL
     * @param params 请求参数
     * @return 服务器响应字符串
     * @throws Exception
     */
    public static String postRequest(String url,Map<String,String> rawParams)
throws Exception{
        HttpPost post=new HttpPost(url);      // 创建 HttpPost 对象
        // 如果传递参数个数比较多的话可以对传递的参数进行封装
        List<NameValuePair> params=new ArrayList<NameValuePair>();
        for(String key : rawParams.keySet()){    //封装请求参数
            params.add(new BasicNameValuePair(key, rawParams.get(key)));
        }
        post.setEntity(new UrlEncodedFormEntity(params, "utf-8"));
                                    //设置请求参数
        HttpResponse httpResponse=httpClient.execute(post);
                                    //发送 POST 请求
        //如果服务器成功地返回响应
        if (httpResponse.getStatusLine().getStatusCode()==200)
        {   // 获取服务器响应字符串
            String result=EntityUtils.toString(httpResponse.getEntity());
            return result;
        }
        return null;
    }
}
```

该程序的运行效果如图 3-17 所示。

图 3-17　获取 Web 网址信息案例运行主界面

3.4 知识扩展

3.4.1 使用 WebView 浏览网页

WebView 组件本身是一个浏览器实现，它的内核基于开源的 WebKit 引擎。

WebView 的用法与普通 ImageView 组件的用法很相似，它常用的函数如下：

```
void goBack();                  //后退
void goForward();               //前进
void loadUrl(String url);       //加载指定 URL 对应的网页
boolean zoomln();               //放大网页
boolean zoomOut();              //缩小网页
```

3.4.2 使用 WebView 中 JavaScript 脚本调用 Android 方法

WebView 加载的页面上常带有 JavaScript 脚本，比如用户单击按钮时弹出提示框或列表框等。在 HTML 页面上的按钮只能激发一段 JavaScript 脚本，这就需要 JavaScript 脚本来调用 Adnroid 的相应方法。在 WebView 的 JavaScript 中调用 Android 方法的步骤如下：

（1）调用 WebView 关联的 WebSettings 的 JavaScript 启用 JavaSrcipt 调用功能。

（2）调用 WebView 的 addJavascriptInterface 方法将 Object 对象暴露给 JavaScript。

（3）在 JavaScript 脚本中通过刚才暴露的 name 对象调用 Android 的方法。

3.5 本章小结

本章主要讲解了 Android 平台上网络与通信的开发，主要包括对无线网络技术的简单介绍，以及网络通信中常见的 HTTP 和 Socket 通信，最后又介绍了 WI-FI 以及蓝牙技术在 Android 中的应用。

3.6 强化练习

1. 填空题

（1）计算机网络是现代通信技术与计算机技术相结合的产物，它是利用（　　）、（　　）和（　　）等网络设备，把分散在各地的计算机连接起来，并通过特定的软件（网络协议）实现计算机之间的相互通信和资源共享。

（2）网络按覆盖范围分类（　　）、（　　）和（　　）。

（3）网络按网络拓扑结构分类（　　）、（　　）和（　　）。

（4）Android 目前有 3 种网络接口可以使用，分别为（　　）、（　　）和（　　）。

（5）HTTP 通信中使用最多的两种方式是（　　）和（　　）。

2. 问答题、操作题或编程题

（1）HTTP Get 与 HTTP Post 两种方式的格式是什么？

（2）如何获得服务器返回的状态码？

（3）在访问网页的时候，权限验证的方法有哪些？

（4）简述 Socket 通信的步骤。

（5）用 Socket 实现简单的通信。

第4章 Android 调用外部数据

学习目标：

- 理解 XML 的概念；
- 掌握 SAX 解析 xml 文件的方法；
- 掌握 DOM 解析 xml 文件的方法；
- 掌握基于位置的服务的实现；
- 掌握基于地图的应用的实现。

4.1 学习导入

XML（Extensible Markup Language，可扩展标记语言）是标准通用标记语言的子集，用于标记电子文件使其具有结构性的标记语言，可以用来标记数据、定义数据类型，是一种允许用户对自己的标记语言进行定义的源语言。它非常适合 Web 传输，提供统一的方法来描述和交换独立于应用程序或供应商的结构化数据。

4.2 技术准备

4.2.1 SAX 解析 xml 文件

SAX（Simple API for XML）既指一种接口，也指一个软件包。SAX 最初是由 David Megginson 采用 Java 语言开发，之后 SAX 很快在 Java 开发者中流行起来。San 现在负责管理其原始 API 的开发工作，这是一种公开的、开放源代码的软件。不同于其他大多数 XML 标准的是，SAX 没有语言开发商必须遵守的标准 SAX 参考版本。因此，SAX 的不同实现可能采用区别很大的接口。

作为接口，SAX 是事件驱动型 XML 解析的一个标准接口（standard interface）不会改变，已被 OASIS（Organization for the Advancement of Structured Information Standards，结构化信息标准促进组织）所采纳。作为软件包，SAX 最早的开发始于 1997 年 12 月，由一些在互联网上分散的程序员合作进行。后来，参与开发的程序员越来越多，组成了互联网上的 XML-DEV 社区。5 个月以后，1998 年 5 月，SAX 1.0 版由 XML-DEV 正式发布。目前，最新的版本是 SAX 2.0。2.0 版本在多处与 1.0 版本不兼容，包括一些类和方法的名字。

由前面的介绍可知，SAX 是一种事件驱动的接口，它的基本原理是由接口的用户提供符合定义的处理器，XML 分析时遇到特定的事件，就去调用处理器中特定事件的处理函数。一般 SAX 接口都是用 Java 实现的，但事实上 C++也可以用于实现 SAX 接口，只是 C++的分析器比较少。之所以称为“简单”应用程序接口，是因为这个接口确实非常简单，绝大多数事情分析器都没有做，需要应用程序自己去实现，因而开发者的任务也相应重一些。

SAX 解析器框架如图 4-1 所示。

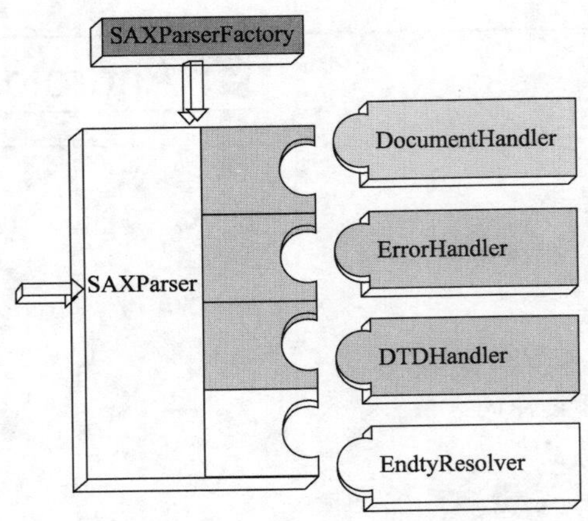

图 4-1　SAX 解析器框架

图中最上方的 SAXParserFactory 用来生成一个分析器实例。XML 文档是从左侧箭头所示处读入，当分析器对文档进行分析时，就会触发在 DocumentHandler、ErrorHandler、DTDHandler 以及 EntityResolver 接口中定义的回调方法。

下面就对 SAX 分析器中的几个主要 API 接口进行简单介绍。

1. ContentHandler 接口

ContentHandler 是 Java 类包中一个特殊的 SAX 接口，位于 org.xml.sax 包中。该接口封装了一些对事件处理的方法，当 XML 解析器开始解析 XML 输入文档时，它会遇到某些特殊的事件，比如文档的开头和结束、元素开头和结束以及元素中的字符数据等事件。当遇到这些事件时，XML 解析器会调用 ContentHandler 接口中相应的方法来响应该事件。

ContentHandler 接口的方法有以下几种：

- void startDocument()
- void endDocument()
- void startElement(String uri, String localName, String qName, Attributes atts)
- void endElement(String uri, String localName, String qName)
- void characters(char[ ] ch, int start, int length)

2. DTDHandler 接口

DTDHandler 用于接收基本的 DTD（Document Type Definition，文档类型定义）相关事件的通知。该接口位于 org.xml.sax 包中。此接口仅包括 DTD 事件的注释和未解析的实体声明部分。SAX 解析器可按任何顺序报告这些事件，而不管声明注释和未解析实体时所采用的顺序；但是，必须在文档处理程序的 startDocument()事件之后，在第一个 startElement()事件之前报告所有的 DTD 事件。

DTDHandler 接口包括以下两个方法：

- void startDocumevoid notationDecl(String name, String publicId, String systemId) nt()

- void unparsedEntityDecl(String name, String publicId, String systemId, String notationName)

3. EntityResolver 接口

EntityResolver 接口是用于解析实体的基本接口，该接口位于 org.xml.sax 包中。该接口只有如下一个方法：

public InputSource resolveEntity(String publicId, String systemId)

解析器将在打开任何外部实体前调用此方法。此类实体包括在 DTD 内引用的外部 DTD 子集和外部参数实体和在文档元素内引用的外部通用实体等。如果 SAX 应用程序需要实现自定义处理外部实体，则必须实现此接口。

4. ErrorHandler 接口

ErrorHandler 接口是 SAX 错误处理程序的基本接口。如果 SAX 应用程序需要实现自定义的错误处理，则它必须实现此接口，然后解析器将通过此接口报告所有的错误和警告。

该接口的方法如下：

- void error(SAXParseException exception)
- void fatalError(SAXParseException exception)
- void warning(SAXParseException exception)

一个典型的 SAX 应用程序至少要提供一个 DocumentHandler 接口。一个健壮的 SAX 应用程序还应该提供 ErrorHandler 接口。

4.2.2 DOM 解析 xml 文件

DOM（Document Object Model，文档对象模型）在本质上是一种文档平台。文档对象模型（DOM）是表示文档（如 HTML 和 XML）和访问、操作构成文档的各种元素的应用程序接口（API）。支持 JavaScript 的所有浏览器都支持 DOM。DOM 实际上是一个能够让程序和脚本动态访问和更新文档内容、结构和样式的一种语言平台。它是基于信息层次的，因而 DOM 被认为是基于树或基于对象的。一方面，DOM 及其广义的基于树的处理具有几个优点。由于树在内存中是持久的，因此可以修改它以便应用程序能够对数据和结构作出更改。它还可以在任何时候在树中上下导航，而不是像 SAX 那样一次性处理。DOM 使用起来也要简单得多。另一方面，DOM 还提供了一个 API，允许开发人员添加、编辑、移动或删除树中任意位置的结点，从而创建一个引用程序。

对于需要自己编写代码来处理 XML 文档的开发人员来说，选择 DOM 还是 SAX 解析模型是一个非常重要的设计决策。DOM 采用树形结构的方式访问 XML 文档，而 SAX 采用事件模型方式。下面简单对比一下两种不同的方式：

SAX 分析器在对 XML 文档进行分析时，触发一系列的事件，应用程序通过事件处理函数实现对 XML 文档的访问。由于事件触发本身是有时序性的，因此，SAX 分析器提供的是一种对 XML 文档的顺序访问机制，对于已经分析过的部分，不能再倒回去重新处理。

SAX 分析器在实现时，它只是顺序地检查 XML 文档中的字节流，判断当前字节是 XML 语法中的哪一部分，检查是否符合 XML 语法并触发相应的事件。对于事件处理函数本身，要由应用程序自己来实现。同 DOM 分析器相比，SAX 分析器对 XML 文档的处理缺乏一定的灵活性，然而，对于那些只需要访问 XML 文档中的数据而不对文档进行更改的应用程序来说，

SAX 分析器的效率则更高。由于 SAX 分析器实现简单，对内存要求比较低，因此实现效率比较高同时具有广泛的应用价值。

DOM 分析器通过对 XML 文档的分析，把整个 XML 文档以一棵 DOM 树的形式存放在内存中，应用程序可以随时对 DOM 树中的任何一个部分进行访问与操作，也就是说，通过 DOM 树，应用程序可以对 XML 文档进行随机访问。这种访问方式给应用程序的开发带来了很大的灵活性，它可以任意地控制整个 XML 文档中的内容。然而，由于 DOM 分析器把整个 XML 文档转化成 DOM 树放在了内存中，因此，当 XML 文档比较大或者文档结构比较复杂时，对内存的需求就比较高。而且，对于结构复杂的树的遍历也是一项比较耗时的操作。所以，DOM 分析器对机器性能的要求比较高，实现效率不十分理想。不过，由于 DOM 分析器的树结构的思想与 XML 文档的结构相吻合，而且，通过 DOM 树机制很容易实现随机访问。因此 DOM 分析器也有较为广泛的使用价值。DOM 分析器的树结构如图 4-2 所示。

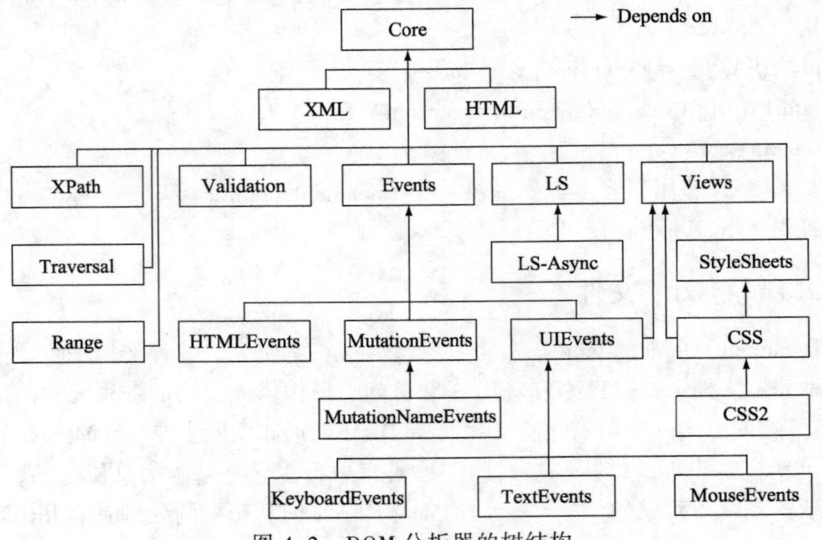

图 4-2　DOM 分析器的树结构

1. SAX

优点：

- 无须将整个文档加载到内存，因而内存消耗少。
- 该模型允许注册多个 ContentHandler。

缺点：

- 没有内置的文档导航支持。
- 不能够随机访问 XML 文档。
- 不支持在原地修改 XML。
- 不支持名字空间作用域。

最适合于：只从 XML 读取数据的应用程序（不可用于操作或修改 XML 文档）。

2. DOM

优点：

- 易于使用。

- 丰富的 API 集合，可用于轻松导航。
- 整棵树加载到内存，允许对 XML 文档进行随机访问。

缺点：

- 整个 XML 文档必须一次解析完。
- 将整棵树加载到内存成本较高。
- 一般的 DOM 结点对必须为所有结点创建对象的对象类型绑定不太理想。

最适合于需要修改 XML 文档的应用程序或 XSLT 应用程序（不可用于只读 XML 的应用程序）。

4.2.3 基于位置的服务

所谓基于位置服务（Location-Based Services，LBS）又称定位服务或基于位置的服务，融合了GPS 定位、移动通信、导航等多种技术，提供了与空间位置相关的综合应用服务。它包括两层含义：首先是确定移动设备或用户所在的地理位置；其次是提供与位置相关的各类信息服务。意指与定位相关的各类服务系统，简称“定位服务”。Android 平台支持提供位置服务的 API，在开发过程中主要用到 LocationManager 和 LocationProviders 对象。LocationManager 可以用来获取当前的位置，追踪设备的移动路线，或设定敏感区域，在进入或离开敏感区域时设备会发出特定警报。LocationProvider 是能够提供定位功能的组件集合，集合中的每种组件以不同的技术提供设备的当前位置，区别在于定位的精度、速度和成本等方面。在 LocationManager 中包含了两种不同的定位技术：GPS 定位技术 GPS_PROVIDER 和网络定位技术 NETWORK_PROVIDER。LocationManager 是用于周期性获得当前设备位置信息的类。要获得 LocationManager 的实例，需要调用 Context.getSystemService()方法，并传入服务名，这里要用到的服务就是 LOCATION_SERVICE，这样就获得了 LocationManager 的一个实例。然后通过 LocationManager 获得要使用的定位技术，若使用 GPS 定位技术则为 LocationManager.GPS_PROVIDER;若使用网络定位则为 LocationManager.NETWORK_PROVIDER。然后通过 getLastKnownLocation()方法将上一次 LocationManager 获得的有效位置信息以 Location 的对象形式返回。具体使用方法如下面的代码段所示：

```
String serviceString=Context.LOCATION_SERVICE;      //获取服务
LocationManager locationManager=(LocationManager)getSystemService (serviceString);
String provider=LocationManager.GPS_PROVIDER;       //通过 GPS 获得位置信息
Location location=locationManager.getLastKnownLocation(provider);
                                                    //获得 location 对象
```

在上面的代码中获取到的服务是 LOCATION_SERVICE，在 Android 系统中有不同的服务，表 4-1 中列出几个常用的系统级服务。

表 4-1 Android 系统常用系统级服务

Context 类的静态常量	值	返回对象	说明
LOCATION_SERVICE	location	LocationManager	控制位置等设备的更新
WINDOW_SERVICE	window	WindowManager	最顶层的窗口管理器
LAYOUT_INFLATER_SERVICE	layout_inflater	LayoutInflater	将 XML 资源实例化为 View
POWER_SERVICE	power	PowerManager	电源管理
ALARM_SERVICE	alarm	AlarmManager	在指定时间接受 Intent

续表

Context 类的静态常量	值	返回对象	说明
NOTIFICATION_SERVICE	notification	NotificationManager	后台事件通知
KEYGUARD_SERVICE	keyguard	KeyguardManager	锁定或解锁键盘
SEARCH_SERVICE	search	SearchManager	访问系统的搜索服务
VIBRATOR_SERVICE	vibrator	Vibrator	访问支持振动的硬件
CONNECTIVITY_SERVICE	connection	ConnectivityManager	网络连接管理
WIFI_SERVICE	wifi	WifiManager	Wi-Fi 连接管理
INPUT_METHOD_SERVICE	input_method	InputMethodManager	输入法管理

通过上面可以获得所在的地理位置，但是由于是处于不断移动状态中，如果 GPS 信息变化了，它不能及时反映到界面上，除非重启程序才能获得新的 GPS 信息，显然这种方法是相当笨拙且不能满足需要的。

要完成上面的需求，这时候就需要启动 LocationManager 的监听器，通过调用 requestLocationUpdates() 方法为其设置一个 LocationListener（位置监听器）。同时 requestLocationUpdates()方法还需要指定要使用的位置服务类型以及位置更新的时间和最小位移，以确保在满足用户需求的前提下最低的耗电量。

requestLocationUpdates(provider, minTime, minDistance, listener)，其中 provider 为提供服务的类型，minTime 为更新的最小间隔，minDistance 为最小的位移变化，listener 为监听方法，为 LocationListener 对象。

4.2.4 基于地图的应用

在 Android SDK1.5 中，以 JAR 包的形式（maps.jar）提供了与 Google Map 相关的 API，来方便开发人员进行地图相关的开发，该 Jar 库位于<SDK>\add-ons\addon_google_apis_ google_inc_7\libs 目录下。这里需要注意，当应用 Google Map 做应用程序时，需要指定使用包含有 Google API 的 Target 作为项目构建的目标，如图 4-3 所示。

Project Build Target

Target Name	Vendor	Platform	API...
☐ Android 1.5	Android Open Source Project	1.5	3
☐ Android 1.6	Android Open Source Project	1.6	4
☐ Android 2.0	Android Open Source Project	2.0	5
☐ Android 2.1	Android Open Source Project	2.1	7
☐ Android 2.2	Android Open Source Project	2.2	8
☐ Google APIs	Google Inc.	2.2	8
☐ Android 2.3.3	Android Open Source Project	2.3.3	10
☐ Android 3.0	Android Open Source Project	3.0	11
☐ Android 3.1	Android Open Source Project	3.1	12
☐ Android 3.2	Android Open Source Project	3.2	13
☐ Android 4.0	Android Open Source Project	4.0	14
☐ Android 4.0.3	Android Open Source Project	4.0.3	15
☐ Google APIs	Google Inc.	4.0.3	15
☐ Android 4.1.2	Android Open Source Project	4.1.2	16
☐ Google APIs	Google Inc.	4.1.2	16
☐ Android 4.2.2	Android Open Source Project	4.2.2	17
☐ Google APIs	Google Inc.	4.2.2	17
☐ Android 4.3	Android Open Source Project	4.3	18
☑ Google APIs	Google Inc.	4.3	18

Standard Android platform 4.0.3

图 4-3 选择 Google API 作为 Target

另外需要注意的是在对程序进行调试的时候必须在带有 Google API 动态链接库的 Android 系统的 AVD 或者真机上运行。否则在安装应用的时候会因为没有找到相应的动态链接库而导致安装不成功。

下面讲解在应用 Google Map 的 API 时在应用中嵌入地图信息，首先向 Google 申请一组经过验证的"地图密钥"（Map API Key），然后使用 MapView（com.google.android.maps.MapView）就可以将 Google 地图嵌入到 Android 应用程序中，才能正常使用 Google 的地图服务。"地图密钥"是访问 Google 地图数据的密钥，无论是模拟器还是在真实设备中都需要使用这个密钥。下面详细讲解如何获取 Map API key。

（1）申请一个 Google 账户，也就是 Gmail 电子邮箱，申请地址是 https://www.google.com/accounts/Login。

找到保存 Debug 证书的 keystore 的保存位置，并获取证书的 MD5 散列值。keystore 是一个密码保护的文件，用来存储 Android 提供的用于调试的证书，获取 MD5 散列值的主要目的是为下一步申请"地图密钥"做准备。首先打开 Eclipse，通过 Window→Preferences 打开配置窗体，在 Android→Build 栏中的 Default debug keystore 中可以找到，如图 4-4 所示。

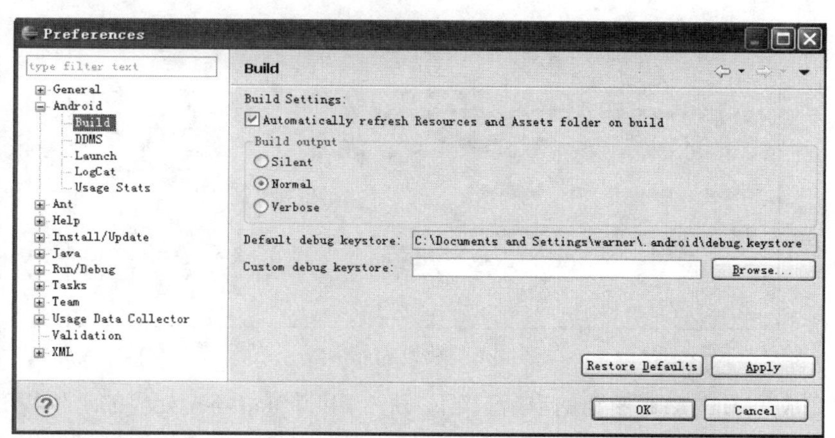

图 4-4　keystore 存放地址

为了获取 Debug 证书 MD5 散列值，需要打开命令行工具 CMD，然后切换到 keystore 目录，输入如下命令，注意，如果提示无法找到 keytool，可以将<Java SDK>/bin 的路径添加到系统的 PATH 变量中：

```
keytool -list -keystore debug.keystore
```

在提示输入 keystore 密码时，输入默认密码 android，MD5 散列将显示在最下方。MD5 散列值为 62:26:AA:22:BE:46:D2:E9:75:6F:B1:77:E1:24:E1:8F。

（2）在有了 MD5 散列值以后，需要打开申请页面，输入散列值，申请地址为：http://code.google.com/intl/zh-CN/android/add-ons/google-apis/maps-api-signup.html。在申请页面中填入 MD5 散列值，然后选择"同意 Android Map API 服务条款"复选框，单击 Generate API key 按钮，即可生成 Map API 密钥，如图 4-5 所示。

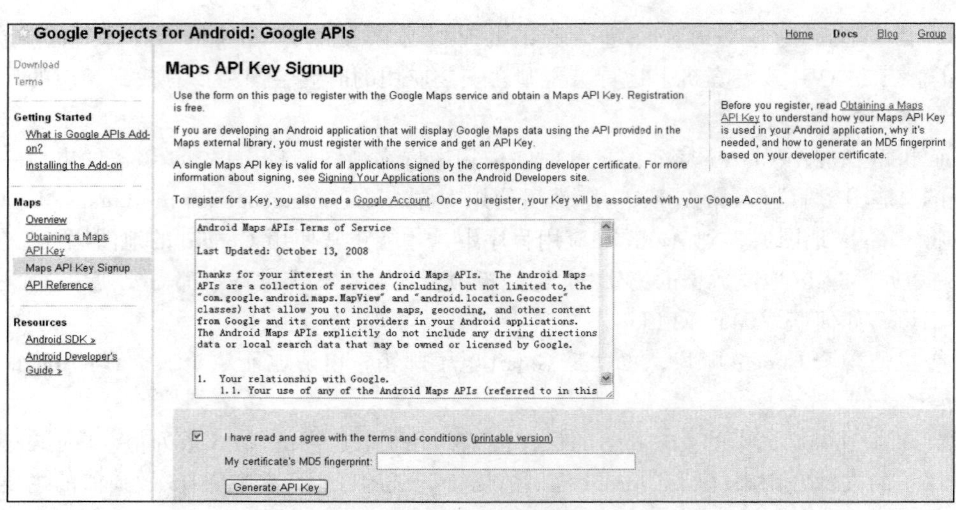

图 4-5　接受服务条款

（3）填入 Google Gmail 账号即可领取 Map API，如图 4-6 所示。

图 4-6　获取 Map API

到此就完成了 Map API 密钥的申请，接下来所有用到 MapView 控件的示例程序中都会使用到它。不过需要注意的是，在应用程序发布时，需要根据为应用程序签名的密钥重新生成 Map API 密钥，并在程序中修改应用到 Map API 密钥的地方。

1．使用 MapView 下载显示地图

为了在应用中使用地图，应用必须从 MapActivity 类继承创建一个新的 Activity，而不是一直习惯用的 Activity 类，然后在 Activity 中加入 MapView，以便绘制 Google 地图。MapView 的基本用法是通过覆盖 onCreate()方法，把 MapView 绘制到屏幕上，同时覆盖 isRouteDisplayed()方法，它表示是否需要在地图上绘制导航线路。在这里会用到两个类，分别为 MapView 类和 MapController 类。MapView 类的主要 API 见表 4-2 所示。MapController 类的主要 API 见表 4-3 所示。

表 4-2　MapView 类的 API 列表

方　法	说　明
canCoverCenter()	检查当前是否有地图贴片覆盖地图中心点
checkLayoutParams (android.view.ViewGroup.LayoutParams p)	仅检查 p 是否为一个 MapView.LayoutParams 实例

续表

方　　法	说　　明
computeScroll()	捕获滚动事件，用它们去平移地图
displayZoomControls(boolean takeFocus)	显示缩放控件，可以选择是否请求焦点选中以便通过按键访问
generateDefaultLayoutParams()	返回一个 Layout 参数的集合，其中参数带有 ViewGroup.LayoutParams.WRAP_CONTENT 的宽度，ViewGroup.LayoutParams.WRAP_CONTENT 高度和坐标(0,0)
getController()	返回地图的 MapController，这个对象可用于控制和驱动平移和缩放
setTraffic(boolean on)	设置为交通视图
setStreetView(boolean on)	设置为街景视图
setSatellite(boolean on)	设置为卫星视图

表 4-3　MapController 类的 API 列表

方　　法	说　　明
animateTo(GeoPoint point)	对已给定的点 GeoPoint，开始动画显示地图
onKey(View v, int keyCode, KeyEvent event)	处理按键事件，把事件变换为适度的地图平移
scrollBy(int x, int y)	按照给定的像素数据量滚动
setCenter(GeoPoint point)	在给定的中心点 GeoPoint 上设置地图视图
setZoom(int zoomLevel)	设置地图的缩放级别
stopAnimation(boolean jumpToFinish)	终止所有未完成的动画，有条件的把地图中心修正到已完成的特殊动画的偏移量上去
stopPanning()	重新设置平移状态，使地图静止
zoomIn()	放大一个级别
zoomInFixing(int xPixel, int yPixel)	放大一个级别；该放大会平移地图使之保持在屏幕的一个固定点上；通过像素坐标（xPixel，yPixel）来设定固定点
zoomToSpan(int latSpanE6, int lonSpanE6)	尝试调整地图的缩放，以便显示给定的经纬度范围

下面通过一个示例简单演示 MapView 的使用方法。首先，需要创建一个项目 MapViewTest，注意 Target 为 Google API，且需要继承自 MapActivity。代码清单：

```
public class MapViewTest extends MapActivity {
  private MapView mapView;
  private MapController mapController;
  public void onCreate(Bundle savedInstanceState) {
     /** 第一次创建活动（activity ）时调用*/
     super.onCreate(savedInstanceState);
     setContentView(R.layout.main);
     mapView=(MapView)findViewById(R.id.mapview);
     mapController=mapView.getController();
     Double lng=126.676530486 * 1E6;
     Double lat=45.7698895661 * 1E6;
     GeoPoint point=new GeoPoint(lat.intValue(), lng.intValue());
      mapController.setCenter(point);
     mapController.setZoom(11);
```

```
        mapController.animateTo(point);
        mapView.setSatellite(false);
    }
    protected boolean isRouteDisplayed() {
        return false;
    }
}
```

还需要注意的是在程序中还用到了一个类，那就是 GeoPoint 类。这个类比较简单，它用来表示一个地理坐标点，存放经度和纬度，以纬度的整数形式存储。所以在程序中需要使用 GeoPoint，通过设定的经纬度得到想要的点。还需要在配置文件 AndroidManifest 中设置使用权限，因为下载地图需要连接网络，所以需要加入对互联网的使用权限。

```
<uses-permission android:name="android.permission.INTERNET"/>
```

又由于 Android 地图相关的 API 包不是系统标准包，而是可选包，因此必须显式地在 AndroidManifest.xml 中声明将使用的地图库：

```
<application android:icon="@drawable/icon" android:label="@string/ app_name">
    <activity android:name=". MapViewTest" android:label="@string/app_name">
        <intent-filter>
            <action android:name="android.intent.action.MAIN" />
            <category android:name="android.intent.category.LAUNCHER" />
        </intent-filter>
    </activity>
    <uses-library android:name="com.google.android.maps"></uses-library>
</application>
```

这里需要注意布局文件，由于在程序中使用 MapView，所以需要在布局文件中加入 <MapView>标签，其使用方法在刚申请到 Map API 密钥时已经有提示，需要注意的是，这里需要提供 Map API 密钥。布局文件的代码段如下面所示，只列出 MapView 的声明方法：

```
<com.google.android.maps.MapView
    android:id="@+id/mapview"
    android:layout_width="fill_parent"
    android:layout_height="fill_parent"
    android:enabled="true"
    android:clickable="true"
    android:apiKey="0SuYC_fdZN0SbWOOCgsfpR2VwPbMRrekTgas0OQ"/>
```

一切准备好后，启动程序，在选用模拟器的时候还需要注意，需要专门创建一个支持 Google APIs 的模拟器来对应项目 Target。程序运行结果如图 4-7 所示。

图 4-7　程序运行图

2. 在地图上标记位置

前面已经能够在应用程序中使用 MapView，但是仅仅能显示是远远不够的，需要在地图上做更多的操作，其中最基本的就是在地图上进行标注。需要标注位置以及其他应用服务。

若要在地图上进行标注，就需要用到 Overlay 类。Overlay 类是一种专门用于在地图上用 2D 图像进行标注的类，Overlay 是一个基类，它表示可以显示在地图上方的覆盖 Overlay。添加一个 Overlay 时，从这个基类派生出一个子类，创建一个实例，然后把它加入到一个列表中。这个列表通过调用 MapView.getOverlays()得到。为了允许用户触摸去对齐一个点，子类应当实现 Overlay.Snappable 接口。表 4-4 就是关于 Overlay 类的基本函数及其说明。

表 4-4　Overlay 主要方法

方　法	说　明
draw(Canvas canvas, MapView mapView, boolean shadow)	在地图上绘制 Overlay
draw(Canvas canvas, MapView mapView, boolean shadow, long when)	专门绘制动画 Overlay 的调用
drawAt(Canvas canvas, Drawable drawable, int x, int y, boolean bShadow)	在某个偏移位置画一个 Drawable 的便捷方法
onKeyDown(int keyCode, KeyEvent event, MapView mapView)	处理一个按键被按下的事件
onKeyUp(int keyCode, KeyEvent event, MapView mapView)	处理一个按键放开事件
onTap(GeoPoint p, MapView mapView)	处理一个“点击”事件
onTouchEvent(MotionEvent e, MapView mapView)	处理一个触摸事件
onTrackballEvent(MotionEvent e, MapView mapView)	处理一个轨迹球事件

要为 MapView 添加 Overlay，只需要用 MapView 对象调用 getOverlays()方法获得该 MapView 所有已经添加的 Overlay 对象的 list（List<Overlay>），然后将自定义的 Overlay 对象添加到这个 list 中即可。

首先通过 findViewById()方法获得 MapView 对象的引用，接着调用 getOverlays()方法获得其 Overlay 的列表，最后将构建好的自定义的 Overlay 对象加到 list 中。其示例代码如下所示：

```
Info info=infoList.get(pos);
Location loc=info.location;
Double latitude=loc.getLatitude()*1E6;
Double longitude=loc.getLongitude()*1E6;
GeoPoint geo=new GeoPoint(latitude.intValue(),longitude.intValue());
RedPoint mark=new RedPoint(geo);
List<Overlay> overlays=map.getOverlays();
overlays.clear();
overlays.add(mark);
```

在完成上面的功能后还要完善其缩放功能，这就用到了 MapView 对象，然后通过它来调用 getControl()方法获得一个 MapController 对象。然后对 MapView 进行缩放。首先获得对应的按钮，然后监听其按钮事件，通过 MapController 对象调用 zoomIn()方法和 zoomOut()方法。这里不再赘述，还是通过一个具体的例子来说明。在这个程序中通过一个 listView 来显示所获得的地震信息列表。然后设置长按事件，监听其长按事件从而获得在地图上的信息。由于在前面已经讲了 xml 的解析，这里只介绍关键的地方，如下代码所示：

```
list.setOnItemLongClickListener(new OnItemLongClickListener() {
    public boolean onItemLongClick(AdapterView<?> parent,View v, int pos,long id){
        Info info=infoList.get(pos);
        Location loc=info.location;
        Double latitude=loc.getLatitude()*1E6;
        Double longitude=loc.getLongitude()*1E6;
        GeoPoint geo=new GeoPoint(latitude.intValue(),longitude.intValue());
        RedPoint mark=new RedPoint(geo);
```

```
        List<Overlay> overlays=map.getOverlays();
        overlays.clear();
        overlays.add(mark);
        map.getController().animateTo(geo);
        return true;
    }
});
```

还需要自定义一个继承自 Overlay 的子类，将位置信息输出到 MapView 中，如下代码所示：

```
protected class RedPoint extends Overlay {
    GeoPoint g;
    public  RedPoint(GeoPoint geo) {
        super();
        g=geo;
    }
    public void draw(Canvas canvas, MapView mapView, boolean shadow) {
        if(shadow==false) {
            Projection projection=mapView.getProjection();
            Point p=new Point();
            projection.toPixels(g, p);
            Paint paint=new Paint();
            paint.setARGB(255,255,0,0);
            paint.setAntiAlias(true);
            canvas.drawCircle(p.x, p.y,5,paint);
        }
        super.draw(canvas, mapView, shadow);
    }
}
```

4.3 案例——Web 服务中的 Json 数据解析

在多平台系统开发过程中，为了适应多个平台的数据交换，通常会提供 Web 服务接口供各个终端调用，Web 服务接口一般会返回 xml 或 Json（JavaScript Object Notation，JavaScript 对象符号）格式的字符串，各个终端需要对这些数据进行解析处理，下面的实例是将 Web 服务返回的 Json 字符串数据反序列化成 Java 对象，方便在 Android 开发中使用这些数据。该程序的界面布局如图 4-8 所示，其详细代码请参见配套光盘中的 CH04_1。

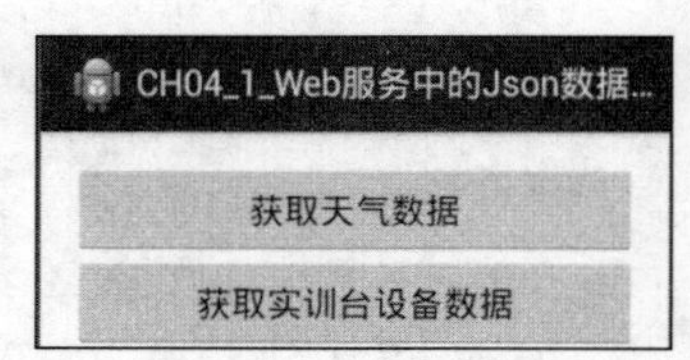

图 4-8　Web 服务中的 Json 数据解析案例布局主界面

注释：Json 是一种轻量级的数据交换格式，其既易于人阅读和编写，也易于机器解析和生成。

程序的界面布局代码如下：CH04\ CH04_1\res\layout\ activity_main.xml。

```
<RelativeLayout xmlns:android="http://schemas.android.com/apk/res/android"
    xmlns:tools="http://schemas.android.com/tools"
    android:layout_width="match_parent"
    android:layout_height="match_parent"
    android:paddingBottom="@dimen/activity_vertical_margin"
```

```
        android:paddingLeft="@dimen/activity_horizontal_margin"
        android:paddingRight="@dimen/activity_horizontal_margin"
        android:paddingTop="@dimen/activity_vertical_margin"
        tools:context=".MainActivity" >
        <Button
            android:id="@+id/btn_getdata_weather"
            android:layout_width="match_parent"
            android:layout_height="wrap_content"
            android:layout_alignParentLeft="true"
            android:layout_alignParentTop="true"
            android:text="@string/getdata_weather" />
        <Button
            android:id="@+id/btn_equipdatas"
            android:layout_width="match_parent"
            android:layout_height="wrap_content"
            android:layout_alignParentLeft="true"
            android:layout_below="@+id/btn_getdata_weather"
            android:text="@string/getdata_equipdatas" />
        <TextView
            android:id="@+id/tv_content"
            android:layout_width="wrap_content"
            android:layout_height="wrap_content"
            android:layout_alignLeft="@+id/btn_getdata_weather"
            android:layout_alignParentBottom="true"
            android:layout_alignParentRight="true"
            android:layout_below="@+id/btn_equipdatas"
            android:scrollHorizontally="true"
            android:text="" />
    </RelativeLayout>
```

程序的 Java 代码如下：CH04\CH04_1\src\com\example\demojson\ MainActivity.java。

```
    public class MainActivity extends Activity {/** * 主页面* */
        private EquipmentListHttpRequestTask mEquipmentListHttpRequestTask;
        protected void onCreate(Bundle savedInstanceState) {
            super.onCreate(savedInstanceState);
            setContentView(R.layout.activity_main);
            //设置获取天气数据的按钮监听器
            //地址: http://m.weather.com.cn/data/101250101.html
           findViewById(R.id.btn_getdata_weather).setOnClickListener(new View.
OnClickListener() {
                public void onClick(View v) {
                    //未实现
                }
            });
            //设置获取实训台设备的数据
            //地址: http://www.iotcase.com:8888//bin/node.cgi?method=get
            findViewById(R.id.btn_equipdatas).setOnClickListener(new View.OnCl
ickListener() {
                public void onClick(View v) {
                    mEquipmentListHttpRequestTask=new EquipmentListHttpRequestTask();
```

```
            mEquipmentListHttpRequestTask.execute();
        }
      });
    }
    private Handler mHandler=new Handler();
    private EquipmentDAO mEquipmentDAO=new EquipmentDAO();
    public boolean onCreateOptionsMenu(Menu menu) {
      getMenuInflater().inflate(R.menu.main, menu);
                                            //膨胀的菜单，将菜单项添加到行动条
      return true;
    }
    /*** 获取 equipment 列表* */
    private class EquipmentListHttpRequestTask extends AsyncTask<String,
Integer, List<EquipmentJson>>{
      protected void onPreExecute() {
        super.onPreExecute();
      }
      protected List<EquipmentJson> doInBackground(String... params) {
        return mEquipmentDAO.getAll();
      }
      protected void onProgressUpdate(Integer... values) {
        super.onProgressUpdate(values);
      }
      protected void onPostExecute(List<EquipmentJson> result) {
        if(result!=null){
          EquipmentJson equipmentJson=null;
          StringBuilder strBuilder=new StringBuilder();
          for(int i=0;i<result.size();i++){
            equipmentJson= result.get(i);
            strBuilder.append("设备编号:"+equipmentJson.nodeId).append ("\r\n");
            strBuilder.append("设备类型:"+equipmentJson.nodeTypeId).append ("\r\n");
            String dataStr="";
            for(int j=0;j<equipmentJson.appData.equipDatas.size();j++){
                dataStr+=String.format("data%s:%s;",equipmentJson.appData.
equipDatas.get(j).index,equipmentJson.appData.equipDatas.get(j).dataValue );
            }
            strBuilder.append("设备数据:"+dataStr).append("\r\n");
            strBuilder.append("更新时间:"+equipmentJson.currentTime).
append ("\r\n");
            setText(strBuilder.toString());
          }
        }
      }
    };
    private void setText(final String str){ /*** 设置文本框内容 * */
      final TextView tv_content=(TextView)findViewById(R.id.tv_content);
      mHandler.post(new Runnable() {
        public void run() {
          tv_content.setText(str);
        }
      });
```

```
    }
}
```

程序清单 CH04\ CH04_1\com\example\demojson\ HttpUtil.java

```
public class HttpUtil /*** Http 封装类*/
{
   public static HttpClient httpClient=new DefaultHttpClient();
   /***创建 HttpClient 对象* */
   /** *
    * @param url 发送请求的 URL
    * @return 服务器响应字符串
    * @throws Exception
    */
   public static String getRequest(String url)
      throws Exception{
      HttpGet get=new HttpGet(url);              //创建 HttpGet 对象
      HttpResponse httpResponse=httpClient.execute(get); //发送 Get 请求
      //如果服务器成功地返回响应
      if(httpResponse.getStatusLine().getStatusCode()==200){
         //获取服务器响应字符串
         String result=EntityUtils.toString(httpResponse.getEntity());
         return result;
      }
      return null;
   }
   /**
    * @param url 发送请求的 URL
    * @param params 请求参数
    * @return 服务器响应字符串
    * @throws Exception
    */
   public static String postRequest(String url,Map<String,String> rawParams)
throws Exception{
      HttpPost post=new HttpPost(url);      //创建 HttpPost 对象
      // 如果传递参数个数比较多的话可以对传递的参数进行封装
      List<NameValuePair> params=new ArrayList<NameValuePair>();
      for(String key : rawParams.keySet()){    //封装请求参数
         params.add(new BasicNameValuePair(key, rawParams.get(key)));
      }
      post.setEntity(new UrlEncodedFormEntity(params, "utf-8"));
                                               //设置请求参数
      HttpResponse httpResponse=httpClient.execute(post);
                                               //发送 Post 请求
      //如果服务器成功返回响应
      if (httpResponse.getStatusLine().getStatusCode()==200)
      {  //获取服务器响应字符串
         String result=EntityUtils.toString(httpResponse.getEntity());
         return result;
      }
      return null;
   }
}
```

程序清单 CH04\ CH04_1\com\example\biz\BaseDAO.java。

```
/*** http://tool.oschina.net/codeformat/json * */
public class BaseDAO {
    /*** 服务器地址* */
    protected static String GATEWAY_BASE_URL="http://192.168.1.201";
    protected ObjectMapper mObjectMapper=new ObjectMapper();
    protected Activity mActivity;
    public BaseDAO(){};
    public BaseDAO(Activity activity)
    {
        mActivity=activity;
    }
}
```

程序清单 CH04\ CH04_1\com\example\biz\EquipmentDAO.java。

```
public class EquipmentDAO extends BaseDAO{ /*** 设备业务处理类* */
    public List<EquipmentJson> getAll(){ /*** 获取所有设备* */
        List<EquipmentJson> equipmentEntityList=null;
        try {
          String result =HttpUtil.getRequest(GATEWAY_BASE_URL+"/bin/node.cgi?
method=get");
           equipmentEntityList=mObjectMapper.readValue(result,new TypeReference
<List<EquipmentJson>>() {
           });
        } catch (JsonParseException e) {
           e.printStackTrace();
        } catch (JsonMappingException e) {
           e.printStackTrace();
        } catch (IOException e) {
           e.printStackTrace();
        } catch (Exception e) {
           e.printStackTrace();
        }
        return equipmentEntityList;
    }
    public EquipmentJson getByNodeId(String nodeId){ /*** 根据 Id 获取设备* */
        EquipmentJson equipmentEntity;
        try {
           String  result=HttpUtil.getRequest(GATEWAY_BASE_URL+"/bin/node.
cgi?method=getByNodeId&nodeId="+nodeId);
           equipmentEntity=mObjectMapper.readValue(result,new TypeReference
<EquipmentJson>() {
           });
           return equipmentEntity;
        } catch (JsonParseException e) {
           e.printStackTrace();
        } catch (JsonMappingException e) {
           e.printStackTrace();
        } catch (IOException e) {
           e.printStackTrace();
```

```
        } catch (Exception e) {
            e.printStackTrace();
        }
        return null;
    }
    /** *发送控制指令* */
    public ControlResultJson sendCmd(String nodeId,String stateFlag,String
stateValue){
        ControlResultJson controlResultJson;
        try {
            String result=HttpUtil.getRequest(GATEWAY_BASE_URL+"/bin/node.cgi?
method=sendCmd&nodeId="+nodeId+"&stateFlag="+stateFlag+"&stateValue="+stateValue);
            controlResultJson=mObjectMapper.readValue(result,new TypeReference
<ControlResultJson>() {
            });
            return controlResultJson;
        } catch (JsonParseException e) {
            e.printStackTrace();
        } catch (JsonMappingException e) {
            e.printStackTrace();
        } catch (IOException e) {
            e.printStackTrace();
        } catch (Exception e) {
            e.printStackTrace();
        }
        return null;
    }
}
```

程序清单：CH04\ CH04_1\com\example\entity\EquipmentJson.java。

```
public class EquipmentJson {/** * 设备信息* */
    public String nodeId;
    public String nodeTypeId;
    public int nodeStatus;
    public String currentTime;
    public AppDataJson appData;
}
```

程序清单：CH04\ CH04_1\com\example\entity\AppDataJson.java。

```
public class AppDataJson{/*** 应用数据* */
    public String appNum;
    public String equipType;
    public String equipNum;
    public String equipPAN;
    public String srcAddress;
    public int hopNum;
    public String equipStatus;
    public List<EquipmentDataJson> equipDatas;
}
```

程序清单：CH04\ CH04_1\com\example\entity\EquipmentDataJson.java。

```
package com.example.entity;
```

```
public class EquipmentDataJson {/*** 设备负载数据* */
    public int valueType;
    public int index;
    public String dataValue;
}
```

该程序的运行效果如图 4-9 所示。

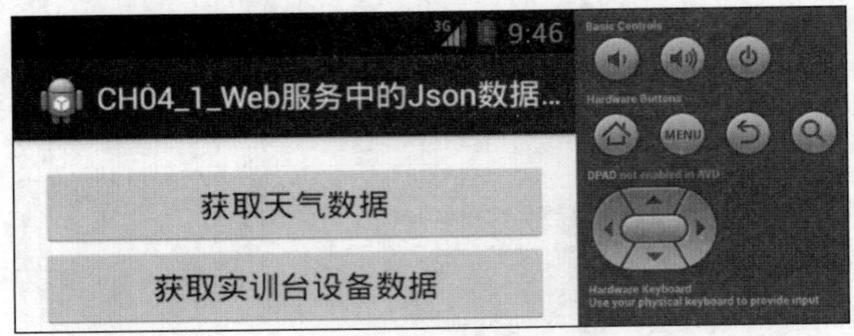

图 4-9　Web 服务中的 Json 数据解析案例运行主界面

4.4 知识扩展

4.4.1 根据 GPS 信息在地图上定位

在 Android 平台上调用 Google Map 服务，Google Map 插件提供了一个 MapView，这个 MapView 的用法就像普通的 ImageView 一样，直接在界面布局文件中定义它，再在程序中通过方法来控制该组件即可。

MapView 支持的常用方法如下：

- MapContrller getController()：获取该 MapView 关联的 MapController。
- GeoPoint getMapCenter()：获取该 MapView 所显示的中心。
- Int getMaxZoomLevel()：获取该 MapView 所支持的最大的放大级别。
- List<Overlay> getOverlays()：获取该 MapView 上显示的全部 Overlay。
- Projection　getProjection()：获取屏幕像素坐标与经纬度坐标之间的投影关系。
- int getZoomLevel()：获取该屏幕当前的缩放级别。
- setBuiltInZoomControls：设置是否显示内置的缩放控制按钮。
- setSatellite：设置是否显示卫星地图。
- setTraffic：设置是否显示交通情况。

4.4.2 调用 Google 的地址解析服务

Google Map 的地图定位必须根据经度和纬度来完成，如果需要让程序根据地址进行定位，需要先把地址解析成经度和纬度。

地址解析：把普通用户能看到的字符串地址转换成经度和纬度。

反向地址解析：把经度和纬度转换成普通的字符串地址。

在 Android 中地址解析通过 Geocoder 工具类实现。该工具类提供了两个方法：

- List<Address> getFromLocation：执行地址解析，把经度和纬度转换为字符串地址值。

- List<Address> getFromLocationName：执行反向地址解析，把字符串地址转换为经度和纬度。

4.5 本章小结

本章主要讲述了 XML 的概念，SAX 解析 xml 文件的方法，DOM 解析 xml 文件的方法，基于位置的服务的实现和基于地图的应用实现。

4.6 强化练习

1. 填空题

（1）XML 非常适合（　　）传输，提供统一的方法来描述和交换独立于应用程序或供应商的结构化数据。

（2）SAX 的全称是（　　），既指一种接口，也指一个软件包。

（3）SAX 是（　　）的一个标准接口。

（4）DOM 的全称是（　　），它在本质上是一种文档平台。

（5）基于位置服务（Location-Based Services，LBS）又称定位服务或基于位置的服务，融合了（　　）、（　　）和（　　）等多种技术，提供了与空间位置相关的综合应用服务。

2. 问答题、操作题或编程题

（1）简述 SAX 分析器中的几个主要 API 接口及其作用。

（2）简述 SAX 和 DOM 各自的优点和不足。

（3）编程实现基于位置的服务。

（4）编程实现基于地图的应用。

第5章 Android 多媒体应用

学习目标：

- 理解 Android 系统中多媒体实现的方法；
- 掌握 Android 中提供的音频相关的类和方法的实现；
- 掌握 Android 中提供的视频相关的类和方法的实现；
- 掌握 Android 中提供的摄像头相关的类和方法的实现。

5.1 学习导入

Android 应用面向的是普通个人用户，这些用户往往更加关注用户体验，因此为 Android 应用增加动画、视频、音乐等多媒体功能十分重要。就目前手持设备的发展趋势来看，手持设备已经不再是单一的通信工具，已经发展成集照相机、视频播放器、个人小型终端于一体的智能设备，因此为手持设备提供音频录制、播放、视频录制、播放的功能十分重要。

5.2 技术准备

5.2.1 使用 MediaPlayer 播放音频

使用 MediaPlayer 播放音频十分简单，当程序控制 MediaPlayer 对象装载音频完成之后，程序可以调用 MediaPlayer 的如下三个方法进行播放控制。

（1）start()：开始或恢复播放。

（2）stop()：停止播放。

（3）pause()：暂停播放。

为了让 MediaPlayer 装载指定音频文件，MediaPlayer 提供了如下简单的静态方法。static MediaPlayer create(Context context,int resid):从 resid 资源 ID 对应的资源文件中装载音频文件，并返回新创建的 MediaPlayer 对象。上面这个方法用起来非常方便，但这个方法每次都会返回新创建的 MediaPlayer 对象，如果程序需要使用 MediaPlayer 循环播放多个音频文件，使用 MediaPlayer 的静态 create 方法就不太适合了，此时可通过 MediaPlayer 的 setDataSource()方法装载指定的音频文件。MediaPlayer 提供了如下方法来指定装载相应的音频文件。

setDataSource(String path)：指定装载 path 路径所代表的文件。

setDataSource(FileDescription fd, long offset, long length)：指定装载 fd 所代表的文件中从 offset 开始、长度为 length 的文件内容。

setDataSource(FileDescription fd)：指定装载 fd 所代表的文件。

setDataSource(Context context,Uri uri)：指定装载 Uri 所代表的文件。

执行上面的 setDataSource()之后，MediaPlayer 并未真正去装载那些音频文件，还需要调用 MediaPlayer 的 prepare()方法去准备音频，所谓“准备”，就是让 MediaPlayer 真正去装载音频文件。因此使用已有的 MediaPlayer 对象装载“下一首”歌曲的代码模板为：

```
try
{
    mPlayer.reset();
    mPlayer.setDatasource("/mnt/sdcard/next.mp3");      //装载下一首歌曲
    mPlayer.prepare();                                   //准备声音
    mPlayer.start();                                     //播放
}
catch(IOException e)
{
    e.printStackTrace();
}
```

除此之外，MediaPlayer 还提供了一些绑定事件监听器的方法，用于监听 MediaPlayer 播放过程中发生的特定事件。绑定事件监听器的方法如下：

setOnCompletionListener(MediaPlayer.OnCompletionListener listener)：为 MediaPlayer 的播放完成事件绑定事件监听器。

setOnErrorListener(MediaPlayer.OnErrorListener listener)：为 MediaPlayer 的播放错误事件绑定事件监听器。

setOnPreparedListener(MediaPlayer.OnPreparedListener listener)：当 MediaPlayer 调用 prepare()方法时触发该监听器。

setOnSeekListener(MediaPlayer.OnSeekCompletedListener listener)：当 MediaPlayer 调用 seek()方法时触发该监听器。

因此可以在创建一个 MediaPlayer 对象之后，通过为该 MediaPlayer 绑定监听器来监听用户所触发的事件。其示例代码如下所示：

```
//为 MediaPlayer 的播放过程中错误事件绑定事件监听器
mPlayer.setOnErrorListener(new OnErrorListener(){
    void onError(MediaPlayer mp, int what, int extra)
    {  //针对错误进行相应的处理
        ...
    }
});
//为 MediaPlayer 的播放完成事件绑定事件监听器
mPlayer.setOnCompletionListener(new OnCompletionListener(){
    public void onCompletion(MediaPlayer mp)
    {
        current++;
        if(current>=3)
        {
            current=0;
        }
        prepareAndPlay(music[current]);
    }
```

```
});
```

下面简单归纳一下 MediaPlayer 播放不同来源的音频文件。

1. 播放应用的资源文件

播放应用的资源文件需要两步即可：

调用 MediaPlayer 对象的 create(Context context, int resid)方法加载指定的资源文件。

调用 MediaPlayer 对象的 start()、pause()、stop()等方法控制播放即可。

其示例代码如下所示：

```
MediaPlayer mPlayer=MediaPlayer.create(this,R.raw.song);
mPlayer.start();
```

2. 播放应用原始文件

播放应用的原始文件按如下步骤执行：

（1）调用 Context 的 getAssets()方法获取应用的 AssetManager。

（2）调用 AssetManager 对象的 openFd(String name)方法打开指定的原始资源，该方法返回一个 AssetFileDescriptor 对象。

（3）调用 AssetFileDescriptor 的 getFileDescriptor()、getStartOffset()和 getLength()方法获取音频文件的 FileDesciptor、开始位置、长度等。

（4）创建 MediaPlayer 对象（或利用已有的 MediaPlayer 对象），并调用 MediaPlayer 对象的 setDataSource(FileDescriptor fd, long offset, long length)方法装载音频资源。

（5）调用 MediaPlayer 对象的 prepare()方法准备音频。

（6）调用 MediaPlayer 对象的 start()、pause()、stop()等方法控制播放即可。

其示例代码如下所示：

```
AssetManager am=getAssets();
AssetFileDescriptor afd=am.openFd(music);  //打开指定音乐文件
MediaPlayer mPlayer=new MediaPlayer();
//使用 MediaPlayer 加载指定的声音文件
mPlayer.setDataSource(afd.getFileDescriptor(),afd.getStartOffset(),afd.
getLength());
mPlayer.prepare();                         //准备声音
mPlayer.start();                           //播放
```

3. 播放外部存储器上的音频文件

播放外部存储器上的音频文件按如下步骤执行：

（1）创建 MediaPlayer 对象（或利用已有的 MediaPlayer 对象），并调用 MediaPlayer 对象的 setDataSource(String path)方法装载指定的音频文件。

（2）调用 MediaPlayer 对象的 prepare()方法准备音频。

（3）调用 MediaPlayer 的 start()、pause()、stop()等方法控制播放即可。

其示例代码如下所示：

```
MediaPlayer mPlayer=new MediaPlayer();
mPlayer.setDataSource("/mnt/sdcard/mysong.mp3");
                                           //使用 MediaPlayer 加载声音文件
mPlayer.prepare();                         //准备声音
mPlayer.start();                           //播放
```

4. 播放来自网络的音频文件

播放来自网络的音频文件有两种方式，第一种，直接使用 MediaPlayer 的静态 create(Context context,Uri uri)方法；第二种，调用 MediaPlayer 的 setDataSource(Context context, Uri uri)装载指定的 Uri 对应的音频文件。

以第二种方式播放来自网络的音频文件的步骤如下：

（1）根据网络上的音频文件所在的位置创建 Uri 对象。

（2）创建 MediaPlayer 对象（或利用已有的 MediaPlayer 对象），并调用 MediaPlayer 对象的 setDataSource(Context context , Uri uri)方法装载 Uri 对应的音频文件。

（3）调用 MediaPlayer 对象的 prepare()方法准备音频。

（4）调用 MediaPlayer 对象的 start()、pause()、stop()等方法控制播放即可。

其示例代码如下所示：

```
Uri uri=Uri.parse("http://www.zig-cloud.com/test.mp3");
MediaPlayer mPlayer=new MediaPlayer();
mPlayer.setDataSource(this,uri);  //使用MediaPlayer根据Uri来加载指定的声音文件
mPlayer.prepare();                //准备声音
mPlayer.start();                  //播放
```

MediaPlayer 除了调用 prepare()方法准备声音之外，还可以调用 prepareAsynchronous()方法准备声音，prepareAsync()方法与普通 prepare()方法的区别在于，prepareAsync()方法是异步的，它不会阻塞当前的 UI 线程。归纳起来，MediaPlayer 状态如图 5-1 所示。

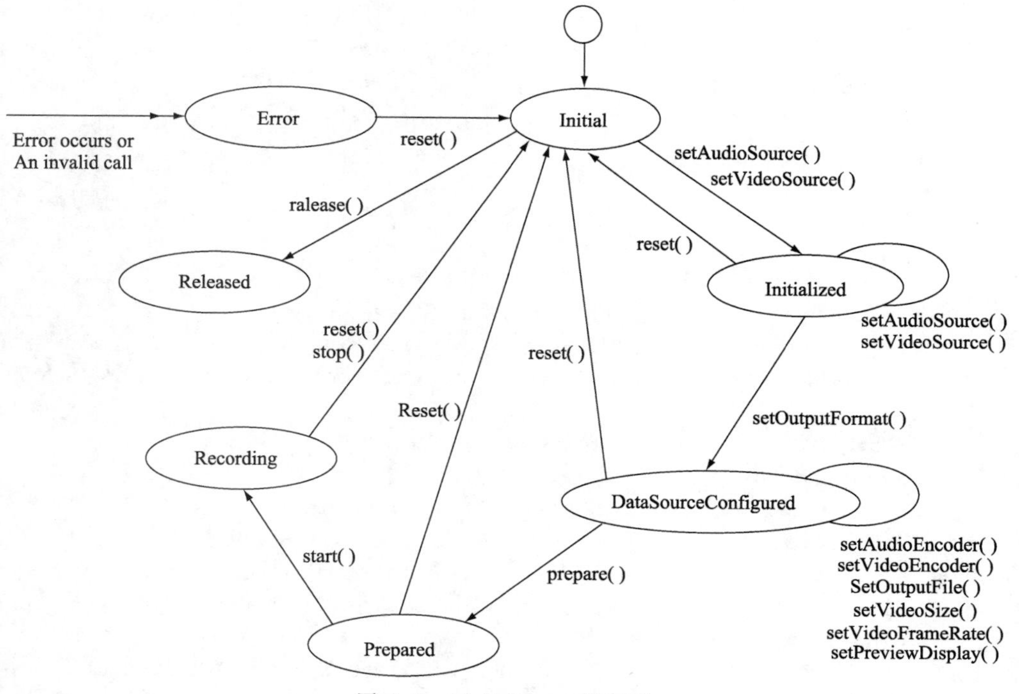

图 5-1　MediaPlayer 状态图

由于前面已经提供了大量使用 MediaPlayer 播放声音的例子，故此处不再介绍使用 MediaPlayer 的简单方法。

MediaPlayer 典型案例：MP3 播放器

该程序实例是无界面布局文件的，使用一个 LinearLayout 容器来装一个示波器 View 组件，来实现频率的均衡值和重低音的强度等功能，其详细代码请参见配套光盘中的 CH05_1。

程序的 Java 代码如下：CH05\ CH05_1\src\com\zigcloud\sound\MediaPlayerTest.java。

```
public class MediaPlayerTest extends Activity
{
   private MediaPlayer mPlayer;              //定义播放声音的 MediaPlayer
   private Visualizer mVisualizer;           //定义系统的示波器
   private Equalizer mEqualizer;             //定义系统的均衡器
   private BassBoost mBass;                  //定义系统的重低音控制器
   private PresetReverb mPresetReverb;   //定义系统的预设音场控制器
   private LinearLayout layout;
   private List<Short> reverbNames=new ArrayList<Short>();
   private List<String> reverbVals=new ArrayList<String>();
   public void onCreate(Bundle savedInstanceState)
   {
      super.onCreate(savedInstanceState);
      setVolumeControlStream(AudioManager.STREAM_MUSIC); //设置控制音乐声音
      layout=new LinearLayout(this);
      layout.setOrientation(LinearLayout.VERTICAL);
      setContentView(layout);
      mPlayer=MediaPlayer.create(this, R.raw.beautiful);
                                             //创建 MediaPlayer 对象
      setupVisualizer();                     //初始化示波器
      setupEqualizer();                      //初始化均衡控制器
      setupBassBoost();                      //初始化重低音控制器
      setupPresetReverb();                   //初始化预设音场控制器
      mPlayer.start();                       //开发播放音乐
   }
   private void setupVisualizer()
   {  //创建 MyVisualizerView 组件，用于显示波形图
      final MyVisualizerView mVisualizerView=new MyVisualizerView(this);
     mVisualizerView.setLayoutParams(new  ViewGroup.LayoutParams(ViewGroup.
LayoutParams.MATCH_PARENT,(int)(120f * getResources().getDisplayMetrics().
density)));
      //将 MyVisualizerView 组件添加到 layout 容器中
      layout.addView(mVisualizerView);
      //以 MediaPlayer 的 AudioSessionId 创建 Visualizer
      //相当于设置 Visualizer 负责显示该 MediaPlayer 的音频数据
      mVisualizer=new Visualizer(mPlayer.getAudioSessionId());
      mVisualizer.setCaptureSize(Visualizer.getCaptureSizeRange()[1]);
      //为 mVisualizer 设置监听器
      mVisualizer.setDataCaptureListener(new Visualizer.OnDataCaptureListener()
         {
            public void onFftDataCapture(Visualizer visualizer,byte[] fft,
int samplingRate)
            {
```

```
                }
                public void onWaveFormDataCapture(Visualizer visualizer,byte[]
waveform, int samplingRate)
                {   //用 waveform 波形数据更新 mVisualizerView 组件
                    mVisualizerView.updateVisualizer(waveform);
                }
            }, Visualizer.getMaxCaptureRate()/2, true, false);
        mVisualizer.setEnabled(true);
    }
    private void setupEqualizer()
    {   //初始化均衡控制器的方法
        //以 MediaPlayer 的 AudioSessionId 创建 Equalizer
        //相当于设置 Equalizer 负责控制该 MediaPlayer
        mEqualizer=new Equalizer(0, mPlayer.getAudioSessionId());
        mEqualizer.setEnabled(true);            //启用均衡控制效果
        TextView eqTitle=new TextView(this);
        eqTitle.setText("均衡器: ");
        layout.addView(eqTitle);
        //获取均衡控制器支持最小值和最大值
        final short minEQLevel=mEqualizer.getBandLevelRange()[0];
        short maxEQLevel=mEqualizer.getBandLevelRange()[1];
        short brands=mEqualizer.getNumberOfBands();
                                                //获取均衡控制器支持的所有频率
        for(short i=0; i<brands; i++)
        {
            TextView eqTextView=new TextView(this);
            //创建一个 TextView，用于显示频率
           eqTextView.setLayoutParams(new ViewGroup.LayoutParams(ViewGroup.
LayoutParams.MATCH_PARENT,ViewGroup.LayoutParams.WRAP_CONTENT));
            eqTextView.setGravity(Gravity.CENTER_HORIZONTAL);
            //设置该均衡控制器的频率
            eqTextView.setText((mEqualizer.getCenterFreq(i)/1000)+" Hz");
            layout.addView(eqTextView);
            //创建一个水平排列组件的 LinearLayout
            LinearLayout tmpLayout=new LinearLayout(this);
            tmpLayout.setOrientation(LinearLayout.HORIZONTAL);
            //创建显示均衡控制器最小值的 TextView
            TextView minDbTextView=new TextView(this);
            minDbTextView.setLayoutParams(new ViewGroup.LayoutParams(ViewGroup.
LayoutParams.WRAP_CONTENT,ViewGroup.LayoutParams.WRAP_CONTENT));
            //显示均衡控制器的最小值
            minDbTextView.setText((minEQLevel/100)+" dB");
            //创建显示均衡控制器最大值的 TextView
            TextView maxDbTextView=new TextView(this);
            maxDbTextView.setLayoutParams(new ViewGroup.LayoutParams(ViewGroup.
LayoutParams.WRAP_CONTENT,ViewGroup.LayoutParams.WRAP_CONTENT));
            maxDbTextView.setText((maxEQLevel/100)+" dB");
                                                    //显示均衡控制器的最大值
```

```
            LinearLayout.LayoutParams layoutParams=new LinearLayout. LayoutParams(
ViewGroup.LayoutParams.MATCH_PARENT,ViewGroup.LayoutParams.WRAP_CONTENT);
            layoutParams.weight=1;
            SeekBar bar=new SeekBar(this);      //定义 SeekBar 做为调整工具
            bar.setLayoutParams(layoutParams);
            bar.setMax(maxEQLevel-minEQLevel);
            bar.setProgress(mEqualizer.getBandLevel(i));
            final short brand=i;
            //为 SeekBar 的拖动事件设置事件监听器
            bar.setOnSeekBarChangeListener(new SeekBar.OnSeekBarChangeListener()
            {
                public void onProgressChanged(SeekBar seekBar,int progress,
boolean fromUser)
                {   //设置该频率的均衡值
                    mEqualizer.setBandLevel(brand,(short) (progress + minEQLevel));
                }
                public void onStartTrackingTouch(SeekBar seekBar)
                {
                }
                public void onStopTrackingTouch(SeekBar seekBar)
                {
                }
            });
            //使用水平排列组件的 LinearLayout“盛装”三个组件
            tmpLayout.addView(minDbTextView);
            tmpLayout.addView(bar);
            tmpLayout.addView(maxDbTextView);
            //将水平排列组件的 LinearLayout 添加到 myLayout 容器中
            layout.addView(tmpLayout);
        }
    }
    // 初始化重低音控制器
    private void setupBassBoost()
    {   //以 MediaPlayer 的 AudioSessionId 创建 BassBoost
        //相当于设置 BassBoost 负责控制该 MediaPlayer
        mBass=new BassBoost(0, mPlayer.getAudioSessionId());
        mBass.setEnabled(true);                //设置启用重低音效果
        TextView bbTitle=new TextView(this);
        bbTitle.setText("重低音: ");
        layout.addView(bbTitle);
        SeekBar bar=new SeekBar(this);  //使用 SeekBar 作为重低音的调整工具
        bar.setMax(1000);                      //重低音的范围为 0~1000
        bar.setProgress(0);
        //为 SeekBar 的拖动事件设置事件监听器
        bar.setOnSeekBarChangeListener(new SeekBar.OnSeekBarChangeListener()
        {
            public void onProgressChanged(SeekBar seekBar,int progress, boolean
fromUser)
```

```
            {
               mBass.setStrength((short) progress);     //设置重低音的强度
            }
            public void onStartTrackingTouch(SeekBar seekBar)
            {
            }
            public void onStopTrackingTouch(SeekBar seekBar)
            {
            }
         });
         layout.addView(bar);
      }
      private void setupPresetReverb()          //初始化预设音场控制器
      {  // 以MediaPlayer的AudioSessionId创建PresetReverb
         // 相当于设置PresetReverb负责控制该MediaPlayer
         mPresetReverb=new PresetReverb(0,mPlayer.getAudioSessionId());
         mPresetReverb.setEnabled(true);        //设置启用预设音场控制
         TextView prTitle=new TextView(this);
         prTitle.setText("音场");
         layout.addView(prTitle);
         for(short i=0; i<mEqualizer.getNumberOfPresets(); i++)
                                                //获取系统支持的所有预设音场
         {
            reverbNames.add(i);
            reverbVals.add(mEqualizer.getPresetName(i));
         }
         Spinner sp=new Spinner(this);          //使用Spinner作为音场选择工具
         sp.setAdapter(new ArrayAdapter<String>(MediaPlayerTest.this,
android.R.layout.simple_spinner_item, reverbVals));
         //为Spinner的列表项选中事件设置监听器
         sp.setOnItemSelectedListener(new Spinner.OnItemSelectedListener()
         {
           public void onItemSelected(AdapterView<?> arg0, View arg1, int arg2,
long arg3)
            {  //设定音场
               mPresetReverb.setPreset(reverbNames.get(arg2));
            }
            public void onNothingSelected(AdapterView<?> arg0)
            {
            }
         });
         layout.addView(sp);
      }
      protected void onPause()
      {
         super.onPause();
         if (isFinishing() && mPlayer != null)
         {  //释放所有对象
```

```
            mVisualizer.release();
            mEqualizer.release();
            mPresetReverb.release();
            mBass.release();
            mPlayer.release();
            mPlayer=null;
        }
    }
    private static class MyVisualizerView extends View
    {   // bytes数组保存了波形抽样点的值
        private byte[] bytes;
        private float[] points;
        private Paint paint=new Paint();
        private Rect rect=new Rect();
        private byte type=0;
        public MyVisualizerView(Context context)
        {
            super(context);
            bytes=null;
            paint.setStrokeWidth(1f);     //设置画笔的属性
            paint.setAntiAlias(true);
            paint.setColor(Color.GREEN);
            paint.setStyle(Style.FILL);
        }
        public void updateVisualizer(byte[] ftt)
        {
            bytes=ftt;
            invalidate();                    //通知该组件重绘自己
        }
        public boolean onTouchEvent(MotionEvent me)
        {   //当用户触碰该组件时，切换波形类型
            if(me.getAction() != MotionEvent.ACTION_DOWN)
            {
                return false;
            }
            type++;
            if(type>=3)
            {
                type=0;
            }
            return true;
        }
        protected void onDraw(Canvas canvas)
        {
            super.onDraw(canvas);
            if(bytes==null)
            {
                return;
```

```
            }
            canvas.drawColor(Color.WHITE); //绘制白色背景（主要为了印刷时好看）
            rect.set(0,0,getWidth(),getHeight());
                                        //使用rect对象记录该组件的宽度和高度
            switch(type)
            {
              case 0:// -------绘制块状的波形图-------
                 for(int i=0; i<bytes.length-1; i++)
                 {
                    float left=getWidth()*i/(bytes.length-1);
                    //根据波形值计算该矩形的高度
                    float top=rect.height()-(byte)(bytes[i+1]+128)*
rect.height()/128;
                    float right=left+1;
                    float bottom=rect.height();
                    canvas.drawRect(left, top, right, bottom, paint);
                 }
                 break;
              case 1: //绘制柱状的波形图（每隔18个抽样点绘制一个矩形）
                 for(int i=0; i<bytes.length-1; i+=18)
                 {
                    float left=rect.width()*i/(bytes.length-1);
                    // 根据波形值计算该矩形的高度
                    float top=rect.height()-(byte)(bytes[i+1]+128)*
rect.height()/128;
                    float right=left+6;
                    float bottom=rect.height();
                    canvas.drawRect(left, top, right, bottom, paint);
                 }
                 break;
              case 2: //绘制曲线波形图
                 if(points==null || points.length<bytes.length*4)
                 {  //如果point数组还未初始化
                    points=new float[bytes.length*4];
                 }
                 for(int i=0; i<bytes.length-1; i++)
                 {  //计算第i个点的x坐标
                    points[i*4]=rect.width()*i/(bytes.length-1);
                    //根据bytes[i]的值（波形点的值）计算第i个点的y坐标
                    points[i*4+1]=(rect.height()/2)+((byte)(bytes[i]+128))*128
/(rect.height()/2);
                    //计算第i+1个点的x坐标
                    points[i*4+2]=rect.width()*(i+1)/(bytes.length-1);
                    //根据bytes[i+1]的值（波形点的值）计算第i+1个点的y坐标
                    points[i*4+3]=(rect.height()/2)+((byte)(bytes[i+1]+128))
*128/(rect.height()/2);
                 }
                 canvas.drawLines(points, paint);      //绘制波形曲线
```

```
                    break;
                }
            }
        }
    }
```

该程序的运行效果如图 5-2 所示。

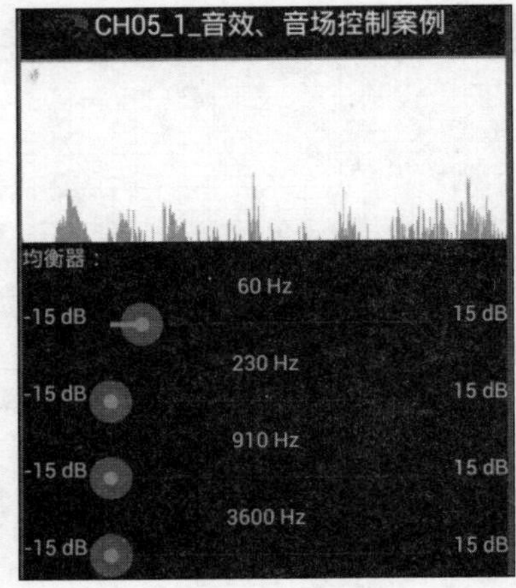

图 5-2　音效、音场案例运行主界面

5.2.2　使用 SoundPool 播放音频

如果应用程序经常需要播放密集、短促的音频，这时 MediaPlayer 就显得有些不合适了。MediaPlayer 存在如下缺点：

（1）资源占用量高、延迟时间较长。

（2）不支持多个音频同时播放。

除了前面介绍的 MediaPlayer 播放音频之外，Android 还提供了 SoundPool 播放音频，SoundPool 使用音频池的概念管理多个短促的音频，例如它可以加载 20 个音频，以后在程序中按音频的 ID 进行播放。

SoundPool 主要用于播放一些较短的声音片段，与 MediaPlayer 相比，SoundPool 的优势在于 CPU 资源占用量低和反应延迟小。另外，SoundPool 还支持执行设置声音的品质、音量、播放比特率等参数。

SoundPool 提供了一个构造器，该构造器可以指定它总共支持多少个声音（也就是池的大小）、声音的品质等。构造器如下：

SoundPool(int maxStreams, int streamType, int srcQuality)：第一个参数指定支持一旦得到了 SoundPool 对象之后，接下来就可调用 SoundPool 的多个重载的 load()方法加载声音，SoundPool 提供了如下 4 个 load()方法。

int load(Context context, int resId,int priority)：从 resId 所对应的资源加载声音。

int load(FileDescriptor fd, long offset, long length, int priority)：加载 fd 所对应的文件的 offset 开始、长度为 length 的声音。

int load(String path, int priority)：从 path 对应的文件加载声音。

int load(AssetFileDescriptor afd, int priority)：从 afd 所在对应的文件中加载声音。

上面三个方法中都有一个 priority 参数，该参数目前还没有任何作用，Android 建议将该参数设置为 1，保持和未来的兼容性。

上面三个方法加载声音之后，都返回该声音的 ID，以后程序就可以通过该声音的 ID 播放指定的声音，SoundPool 提供的播放指定声音的方法如下：

int play(int soundID, float leftVolume, float rightVolume, int priority, int loop, float rate)：该方法的第一个参数指定播放哪个声音；leftVolume、rightVolume 指定左、右的音量；priority 指定播放声音的优先级，数值越大，优先级越高；loop 指定是否循环，0 为不循环，-1 为循环；rate 指定播放的比率，数值范围 0.5～2，为正常比率。

为了更好地管理 SoundPool 加载的每个声音的 ID，程序一般会使用 HashMap<Integer,Integer> 对象管理声音。

归纳起来，使用 SoundPool 播放声音的步骤如下：

（1）调用 SoundPool 的构造器创建 SoundPool 的对象。

（2）调用 SoundPool 对象的 load()方法从指定资源、文件中加载声音。最好使用 HashMap <Integer,Integer>管理加载声音。

（3）调用 SoundPool 的 play 声音播放声音。

SoundPool 典型案例：播放音频

该案例示范了如何使用 SoundPool 播放音频，该程序提供了三个按钮，三个按钮分别用于播放不同的声音。该程序的界面十分简单，故此处不再给出界面布局代码，程序代码如下：

```
public class SoundPoolTest extends Activity implements OnClickListener
{
    Button bomb, shot, arrow;
    SoundPool soundPool;                //定义一个 SoundPool
    HashMap<Integer, Integer> soundMap=new HashMap<Integer, Integer>();
    public void onCreate(Bundle savedInstanceState)
    {
        super.onCreate(savedInstanceState);
        setContentView(R.layout.main);
        bomb=(Button) findViewById(R.id.bomb);
        shot=(Button) findViewById(R.id.shot);
        arrow=(Button) findViewById(R.id.arrow);
        //设置最多可容纳 10 个音频流，音频的品质为 5
        soundPool=new SoundPool(10, AudioManager.STREAM_SYSTEM, 5);
        //load 方法加载指定音频文件，并返回所加载的音频 ID
        //此处使用 HashMap 管理这些音频流
        soundMap.put(1, soundPool.load(this, R.raw.bomb, 1));
        soundMap.put(2, soundPool.load(this, R.raw.shot, 1));
        soundMap.put(3, soundPool.load(this, R.raw.arrow, 1));
        bomb.setOnClickListener(this);
        shot.setOnClickListener(this);
        arrow.setOnClickListener(this);
    }
    public void onClick(View source) //重写 OnClickListener 监听器接口的方法
    {   //判断哪个按钮被单击
        switch(source.getId())
        {
            case R.id.bomb:
                soundPool.play(soundMap.get(1), 1, 1, 0, 0, 1);
                break;
            case R.id.shot:
                soundPool.play(soundMap.get(2), 1, 1, 0, 0, 1);
                break;
            case R.id.arrow:
                soundPool.play(soundMap.get(3), 1, 1, 0, 0, 1);
                break;
        }
    }
}
```

实际使用 SoundPool 播放声音时需要注意：SoundPool 虽然可以一次性加载多个声音，但由于内存限制，因此应该避免使用 SoundPool 来播放歌曲或者做游戏背景音乐，只有那些短促、密集的声音才考虑使用 SoundPool 进行播放。

虽然 SoundPool 比 MediaPlayer 的效率好，但也不是绝对不存在延迟问题，尤其在那些性能不太好的手持设备中，SoundPool 的延迟问题会更严重。该程序运行效果如图 5-3 所示。

图 5-3　播放音效案例运行主界面

5.2.3　使用 VideoView 播放视频

为了在 Android 应用中播放视频，Android 提供了 Video 组件，它就是一个位于 android.widget 包下的组件，它的作用与 ImageView 类似，只是 ImageView 用于显示图片，而 VideoView 用于播放视频。使用 VideoView 播放视频的步骤如下：

（1）在界面布局中定义 VideoView 组件，或在程序中创建 VideoView 组件。

（2）调用 VideoView 对象的如下两个方法来加载指定视频。

- setVideoPath(String path)：加载 path 文件所代表的视频。
- setVideoURI(Uri uri)：加载 uri 所对应的视频。

（3）调用 VideoView 对象的 start()、stop()、pause()方法控制视频播放。

实际上与 VideoView 一起结合使用的还有一个 MediaController 类，它的作用是提供一个友好的图形控制界面，通过控制界面来控制视频的播放。

VideoView 典型案例：播放视频

该案例示范了如何使用 VideoView 播放视频。该程序提供了一个简单的界面，界面布局代码如下：

程序清单：CH05 \CH05_3\res\layout\ activity_main.xml。

```
<?xml version="1.0" encoding="utf-8"?>
<LinearLayout
   xmlns:android="http://schemas.android.com/apk/res/android"
   android:orientation="vertical"
   android:layout_width="match_parent"
   android:layout_height="match_parent">
<!-- 定义 VideoView 播放视频 -->
<VideoView
   android:id="@+id/video"
   android:layout_width="match_parent"
   android:layout_height="match_parent" />
</LinearLayout>
```

上面的布局中定义了一个 VideoView 组件，接下来就可以在程序中使用该组件播放视频。播放视频时还结合了 MediaController 控制视频播放，该程序代码如下：

程序清单：CH05 \CH05_3\src\com\zigcloud\sound\ VedioViewTest.java。

```
public class VedioViewTest extends Activity
{
   VideoView videoView;
   MediaController mController;
```

```
    public void onCreate(Bundle savedInstanceState)
    {
        super.onCreate(savedInstanceState);
        getWindow().setFormat(PixelFormat.TRANSLUCENT);
        setContentView(R.layout.main);
        videoView=(VideoView) findViewById(R.id.video);
                                        //获取界面上 VideoView 组件
        mController=new MediaController(this);
                                        //创建 MediaController 对象
        File video=new File("/mnt/sdcard/movie.mp4");
        if(video.exists())
        {
            videoView.setVideoPath(video.getAbsolutePath());
            // 设置 videoView 与 mController 建立关联
            videoView.setMediaController(mController);
            // 设置 mController 与 videoView 建立关联
            mController.setMediaPlayer(videoView);
            videoView.requestFocus();     // 让 VideoView 获取焦点
        }
    }
}
```

运行该程序，保证/mnt/sdcard/movie.mp4 视频文件存在的前提下，可以正常运行该程序。

5.2.4 使用 MediaPlayer 与 SurfaceView 播放视频

使用 VideoView 播放视频简单、方便，但是早期开发者更喜欢使用 MediaPlayer 播放视频。但由于 MediaPlayer 主要用于播放音频，因此它没有提供图像输出界面，此时就需要借助 SurfaceView 显示 MediaPlayer 播放的图像输出。使用 MediaPlayer 播放视频的步骤如下：

（1）创建 MediaPlayer 对象，并让它加载指定的视频文件。

（2）在界面布局文件中定义 SurfaceView 组件，或在程序中创建 SurfaceView 组件。并为 SurfaceView 的 SurfaceHolder 添加 Callback 监听器。

（3）调用 MediaPlayer 对象的 setDisplay(SurfaceHolder sh)将所播放的视频图像输出到指定的 SurfaceView 组件。

（4）调用 MediaPlayer 对象的 start()、stop()和 pause()方法控制视频播放。

播放视频典型案例：使用 Surface 播放视频

该案例使用了 MediaPlayer 和 SurfaceView 播放视频，并在该程序中使用了三个按钮来控制视频播放、暂停和停止。该程序的代码如下：

程序清单：CH05\ CH05_4\src\com\zigcloud\sound\ SurfaceViewPlayVideo.java。

```
public class SurfaceViewPlayVideo extends Activity implements OnClickListener
{
    SurfaceView surfaceView;
    ImageButton play, pause, stop;
    MediaPlayer mPlayer;
    int position;                              //记录当前视频的播放位置
```

```
public void onCreate(Bundle savedInstanceState)
{
   super.onCreate(savedInstanceState);
   setContentView(R.layout.main);
   //获取界面中的三个按钮
   play=(ImageButton) findViewById(R.id.play);
   pause=(ImageButton) findViewById(R.id.pause);
   stop=(ImageButton) findViewById(R.id.stop);
   //为三个按钮的单击事件绑定事件监听器
   play.setOnClickListener(this);
   pause.setOnClickListener(this);
   stop.setOnClickListener(this);
   mPlayer=new MediaPlayer();        //创建MediaPlayer
   surfaceView=(SurfaceView) this.findViewById(R.id.surfaceView);
   //设置播放时打开屏幕
   surfaceView.getHolder().setKeepScreenOn(true);
   surfaceView.getHolder().addCallback(new SurfaceListener());
}
public void onClick(View source)
{
   try
   {
      switch (source.getId())
      {
         case R.id.play:          // 播放按钮被单击
            play();
            break;
         case R.id.pause:         // 暂停按钮被单击
            if (mPlayer.isPlaying())
            {
               mPlayer.pause();
            }
            else
            {
               mPlayer.start();
            }
            break;
         case R.id.stop:          // 停止按钮被单击
            if (mPlayer.isPlaying()) mPlayer.stop();
            break;
      }
   }
   catch (Exception e)
   {
      e.printStackTrace();
   }
```

```
        }
        private void play() throws IOException
        {
            mPlayer.reset();
            mPlayer.setAudioStreamType(AudioManager.STREAM_MUSIC);
            mPlayer.setDataSource("/mnt/sdcard/movie.3gp");
                                                   //设置需要播放的视频
            mPlayer.setDisplay(surfaceView.getHolder());
                                                   //把视频画面输出到 SurfaceView
            mPlayer.prepare();
            WindowManager wManager=getWindowManager();
                                                   //获取窗口管理器
            DisplayMetrics metrics=new DisplayMetrics();
            wManager.getDefaultDisplay().getMetrics(metrics); //获取屏幕大小
            // 设置视频保持纵横比缩放到占满整个屏幕
            surfaceView.setLayoutParams(new LayoutParams(metrics.widthPixels,
mPlayer.getVideoHeight()*metrics.widthPixels/mPlayer. getVideoWidth()));
            mPlayer.start();
        }
        private class SurfaceListener implements SurfaceHolder.Callback
        {
            public void surfaceChanged(SurfaceHolder holder, int format,
int width, int height)
            {
            }
            public void surfaceCreated(SurfaceHolder holder)
            {
                if(position>0)
                {
                    try
                    {
                        play();                     // 开始播放
                        mPlayer.seekTo(position); //指定开始播放位置
                        position=0;
                    }
                    catch(Exception e)
                    {
                        e.printStackTrace();
                    }
                }
            }
            public void surfaceDestroyed(SurfaceHolder holder)
            {
            }
        }
        protected void onPause()
```

```
    {    //当其他 Activity 被打开，暂停播放
       if (mPlayer.isPlaying())
       {   //保存当前的播放位置
          position=mPlayer.getCurrentPosition();
          mPlayer.stop();
       }
       super.onPause();
    }
    protected void onDestroy()
    {
       if (mPlayer.isPlaying()) mPlayer.stop();          // 停止播放
       mPlayer.release();                                // 释放资源
       super.onDestroy();
    }
}
```

该程序的运行效果如图 5-4 所示。

从上面的开发过程不难看出，使用 MediaPlayer 播放视频要复杂一些，并且需要自己开发控制按钮来控制视频播放。因此一般推荐使用 VideoView 来播放视频。

图 5-4　播放视频案例运行主界面

5.2.5　使用 MediaRecorder 录制音频

手持设备一般都提供了麦克风硬件，而 Android 系统可以利用硬件来录制音频。为了在 Android 系统中录制音频，Android 提供了 MediaRecorder 类。使用 MediaRecorder 录制音频的过程很简单，按照如下步骤进行即可：

（1）创建 MediaRecorder 对象。

（2）调用 MediaRecorder 对象的 setAudioSource()方法设置声音来源，一般传入 MediaRecorder.AudioSource.MIC 参数指定录制来自麦克风的声音。

（3）调用 MediaRecorder 对象的 setOutputFormat()方法设置所录制的音频文件格式。

（4）调用 MediaRecorder 对象的 setAudioEncoder()、setAudioEncodingBitRate(int bitRate)、setAudioSamplingRate(int samplingRate)方法设置所录制的声音的编码格式、编码位率、采样率等，这些参数都可以控制所录制的声音的品质、文件的大小。一般来说，声音品质越好，声音文件越大。

（5）调用 MediaRecorder 对象的 setOutputFile(String path)方法设置录制的音频文件的保存位置。

（6）调用 MediaRecorder 对象的 prepare()方法准备录制。

（7）调用 MediaRecorder 对象的 start()方法开始录制。

（8）录制完成，调用 MediaRecoder 对象的 stop()方法停止录制，并调用 release()方法释放资源。

图 5-5 显示了 MediaRecorder 的状态。

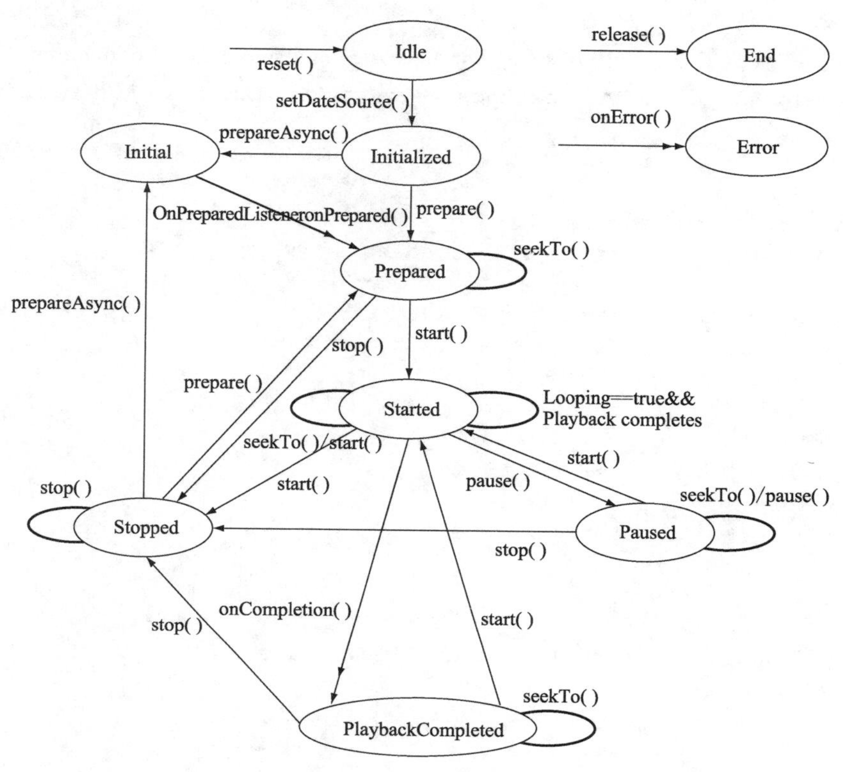

图 5-5　MediaRecorder 的状态图

MediaRecorder 典型案例：迷你录音机

具体的代码请见配套光盘中的 CH05_5。该程序通过 MediaRecorder 实现声音的录制功能，程序的界面布局如图 5-6 所示。

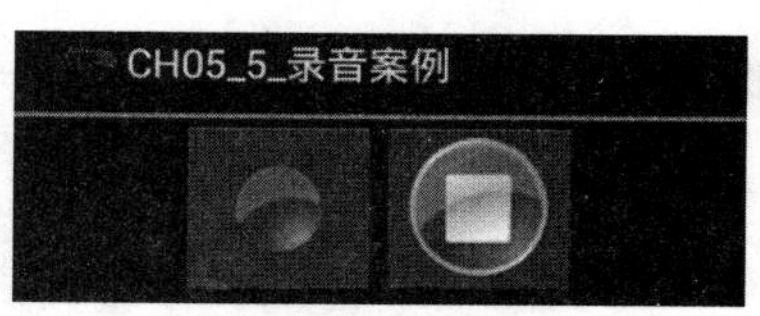

图 5-6　录音案例运行主界面

程序的界面布局文件如下：CH05\CH05_5\res\layout\activity_main.xml。

```
<?xml version="1.0" encoding="utf-8"?>
<LinearLayout xmlns:android="http://schemas.android.com/apk/res/android"
    android:orientation="horizontal"
    android:layout_width="fill_parent"
    android:layout_height="fill_parent"
    android:gravity="center_horizontal" >
<ImageButton
   android:id="@+id/record"
   android:layout_width="wrap_content"
   android:layout_height="wrap_content"
   android:src="@drawable/record" />
<ImageButton
    android:id="@+id/stop"
    android:layout_width="wrap_content"
    android:layout_height="wrap_content"
    android:src="@drawable/stop" />
```

```
</LinearLayout>
```

程序的 Java 代码如下：CH05\CH05_5\src\com\zigcloud\sound\ RecordSound.java

```
public class RecordSound extends Activity implements OnClickListener
{
   ImageButton record, stop;            //定义界面上的两个按钮
   File soundFile;                      //系统的音频文件
   MediaRecorder mRecorder;
   public void onCreate(Bundle savedInstanceState)
   {
      super.onCreate(savedInstanceState);
      setContentView(R.layout.activity_main);
      //获取程序界面中的两个按钮
      record=(ImageButton) findViewById(R.id.record);
      stop=(ImageButton) findViewById(R.id.stop);
      //为两个按钮的单击事件绑定监听器
      record.setOnClickListener(this);
      stop.setOnClickListener(this);
   }
   public void onDestroy()
   {
      if(soundFile != null && soundFile.exists())
      {
         mRecorder.stop();              // 停止录音
         mRecorder.release();           // 释放资源
         mRecorder=null;
      }
      super.onDestroy();
   }
   public void onClick(View source)
   {
      switch (source.getId())
      {
         case R.id.record:              //单击录音按钮
            if (!Environment.getExternalStorageState().equals(
android.os.Environment.MEDIA_MOUNTED))
            {
               Toast.makeText(RecordSound.this, "SD卡不存在，请插入SD卡！",
Toast.LENGTH_SHORT).show();
               return;
            }
            try
            {  //创建保存录音的音频文件
               soundFile=new File(Environment.getExternalStorageDirectory().
getCanonicalFile()+ "/sound.amr");
               mRecorder=new MediaRecorder();
               //设置录音的声音来源
               mRecorder.setAudioSource(MediaRecorder.AudioSource.MIC);
               //设置录制的声音的输出格式（必须在设置声音编码格式之前设置）
               mRecorder.setOutputFormat(MediaRecorder.OutputFormat.THREE_GPP);
               //设置声音编码的格式
```

```
            mRecorder.setAudioEncoder(MediaRecorder.AudioEncoder.AMR_NB);
            mRecorder.setOutputFile(soundFile.getAbsolutePath());
            mRecorder.prepare();
            mRecorder.start();          //开始录音
         }
         catch (Exception e)
         {
            e.printStackTrace();
         }
         break;
      case R.id.stop:                   //单击停止按钮
         if(soundFile != null && soundFile.exists())
         {
            mRecorder.stop();           //停止录音
            mRecorder.release();        //释放资源
            mRecorder=null;
         }
         break;
      }
   }
}
```

该程序的运行效果如图 5-7 所示。

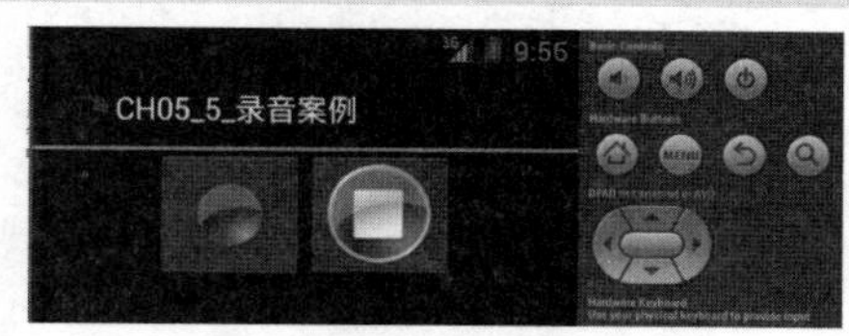

图 5-7 录音案例运行主界面

5.2.6 使用摄像头拍照

现在的手持设备一般都会提供照相机功能，有些照相机的镜头甚至支持 1000 万以上像素，有些甚至支持光学变焦，这些手持设备已经变成了专业的数码照相机。为了充分利用手持设备上的照相机功能，Android 应用可以控制拍照和录制视频。Android 提供了 Camera 控制拍照，使用 Camera 进行拍照也比较简单，按照如下步骤进行即可：

（1）调用 Camera 的 open()方法打开照相机。该方法默认打开后置摄像头。如果需要打开指定的摄像头，可以为该方法传入摄像头 ID。

（2）调用 Camera 的 getParameters()方法获取拍照参数。该方法返回一个 Camera.Parameters 对象。

（3）调用 Camera.Parameters 对象方法设置拍照参数。

（4）调用 Camera 的 startPreview()方法开始预览取景，在预览取景之前需要调用 Camera 的 setPreviewDisplay(SurfaceHolder holder)方法设置使用哪个 SurfaceView 显示取景图片。

（5）调用 Camera 的 takePicture()方法进行拍照。

（6）结束程序时，调用 Camera 的 stopPreview()方法结束取景预览，并调用 release()方法释放资源。

Camera 典型应用案例：迷你照相机

具体的代码请见配套光盘中的 CH05_6。该程序使用 Camera 进行拍照，当用户按下拍照键时，该应用会自动对焦，当对焦成功时拍下照片。该程序的界面布局如图 5-8 所示。

图 5-8 照相机案例布局主界面

程序的界面布局代码如下：CH05\CH05_6\res\layout\ activity_main.xml。

```
<?xml version="1.0" encoding="utf-8"?>
<RelativeLayout xmlns:android="http://schemas.android.com/apk/res/android"
    android:orientation="vertical"
    android:layout_width="match_parent"
    android:layout_height="match_parent">
<SurfaceView
   android:id="@+id/sView"
   android:layout_width="match_parent"
   android:layout_height="match_parent"/>
<ImageButton
   android:layout_width="wrap_content"
   android:layout_height="wrap_content"
   android:onClick="capture"
   android:src="@drawable/capture"
   android:layout_alignParentBottom="true"
   android:layout_alignParentRight="true"/>
</RelativeLayout>
```

程序清单：CH05\CH05_6\res\layout\ save.xml。

```
<?xml version="1.0" encoding="utf-8"?>
<LinearLayout xmlns:android="http://schemas.android.com/apk/res/android"
   android:orientation="vertical"
   android:layout_width="fill_parent"
   android:layout_height="fill_parent">
<LinearLayout
   android:orientation="horizontal"
   android:layout_width="fill_parent"
   android:layout_height="wrap_content">
<TextView
   android:layout_width="wrap_content"
   android:layout_height="wrap_content"
   android:layout_marginRight="8dip"
   android:text="@string/photo_name"/>
<!-- 定义一个文本框让用户输入照片名 -->
<EditText
   android:id="@+id/phone_name"
   android:layout_width="fill_parent"
   android:layout_height="wrap_content"/>
</LinearLayout>
<!-- 定义一个图片框显示照片 -->
<ImageView
   android:id="@+id/show"
   android:layout_width="240px"
   android:layout_height="315px"
   android:scaleType="fitCenter"
   android:layout_marginTop="10dp"/>
</LinearLayout>
```

程序的 Java 代码如下：CH05\CH05_6\src\com\zigcloud\sound\CaptureImage.java。

```
public class CaptureImage extends Activity
```

```
    {
        SurfaceView sView;
        SurfaceHolder surfaceHolder;
        int screenWidth, screenHeight;
        Camera camera;                              //定义系统所用的照相机
        boolean isPreview=false;              //是否在预览中
        public void onCreate(Bundle savedInstanceState)
        {
            super.onCreate(savedInstanceState);
            requestWindowFeature(Window.FEATURE_NO_TITLE);   //设置全屏
            getWindow().setFlags(WindowManager.LayoutParams.FLAG_FULLSCREEN,
WindowManager.LayoutParams.FLAG_FULLSCREEN);
            setContentView(R.layout.activity_main);
            WindowManager wm=getWindowManager();            //获取窗口管理器
            Display display=wm.getDefaultDisplay();
            DisplayMetrics metrics=new DisplayMetrics();
            display.getMetrics(metrics);         //获取屏幕的宽和高
            screenWidth=metrics.widthPixels;
            screenHeight=metrics.heightPixels;
            sView=(SurfaceView) findViewById(R.id.sView);
                                                    //获取界面中 SurfaceView 组件
            //设置该 Surface 不需要自己维护缓冲区
            sView.getHolder().setType(SurfaceHolder.SURFACE_TYPE_PUSH_BUFFERS);
            surfaceHolder=sView.getHolder();//获得 SurfaceView 的 SurfaceHolder
            //为 surfaceHolder 添加一个回调监听器
            surfaceHolder.addCallback(new Callback()
            {
                public void surfaceChanged(SurfaceHolder holder, int format, int width,
int height)
                {
                }
                public void surfaceCreated(SurfaceHolder holder)
                {
                    initCamera();                    // 打开摄像头
                }
                public void surfaceDestroyed(SurfaceHolder holder)
                {  //如果 camera 不为 null，释放摄像头
                    if(camera!=null)
                    {
                        if(isPreview) camera.stopPreview();
                        camera.release();
                        camera=null;
                    }
                }
            });
        }
        private void initCamera()
        {
            if(!isPreview)
            {  //此处默认打开后置摄像头，通过传入参数可以打开前置摄像头
```

```
        camera=Camera.open(0);
        camera.setDisplayOrientation(90);
    }
    if(camera != null && !isPreview)
    {
        try
        {
            Camera.Parameters parameters=camera.getParameters();
            parameters.setPreviewSize(screenWidth, screenHeight);
                                              //设置预览照片的大小
            //设置预览照片时每秒显示多少帧的最小值和最大值
            parameters.setPreviewFpsRange(4, 10);
            parameters.setPictureFormat(ImageFormat.JPEG);  //设置图片格式
            parameters.set("jpeg-quality", 85);  // 设置 JPG 照片的质量
            parameters.setPictureSize(screenWidth, screenHeight);
                                              //设置照片的大小
            camera.setPreviewDisplay(surfaceHolder);
                                              //通过 SurfaceView 显示取景画面
            camera.startPreview();            // 开始预览
        }
        catch(Exception e)
        {
            e.printStackTrace();
        }
        isPreview=true;
    }
}
public void capture(View source)
{
    if(camera != null)
    {  //控制摄像头自动对焦后才拍照
        camera.autoFocus(autoFocusCallback);
    }
}
AutoFocusCallback autoFocusCallback = new AutoFocusCallback()
{  //当自动对焦时激发该方法
    public void onAutoFocus(boolean success, Camera camera)
    {
        if (success)
        {  //takePicture()方法需要传入三个监听器参数
            //第一个监听器：当用户按下快门时激发该监听器
            //第二个监听器：当照相机获取原始照片时激发该监听器
            //第三个监听器：当照相机获取 JPG 照片时激发该监听器
            camera.takePicture(new ShutterCallback()
            {
                public void onShutter()
                {
                    //按下快门瞬间会执行此处代码
                }
            }, new PictureCallback()
```

```
            {
                public void onPictureTaken(byte[] data, Camera c)
                {
                    //此处代码可以决定是否需要保存原始照片信息
                }
            }, myJpegCallback);
        }
    }
};

PictureCallback myJpegCallback=new PictureCallback()
{
    public void onPictureTaken(byte[] data, Camera camera)
    {  //根据拍照所得的数据创建位图
        final Bitmap bm=BitmapFactory.decodeByteArray(data,0,data. length);
        //加载/layout/save.xml文件对应的布局资源
        View saveDialog=getLayoutInflater().inflate(R.layout.save,null);
        final EditText photoName=(EditText)saveDialog.findViewById (R.id.
phone_name);
        //获取saveDialog对话框上的ImageView组件
        ImageView show=(ImageView) saveDialog.findViewById(R.id.show);
        show.setImageBitmap(bm);      // 显示刚刚拍得的照片
        //使用对话框显示saveDialog组件
        new AlertDialog.Builder(CaptureImage.this).setView(saveDialog)
.setPositiveButton("保存", new OnClickListener()
            {
                public void onClick(DialogInterface dialog, int which)
                {    //创建一个位于SD卡上的文件
                    File file=new File(Environment.getExternalStorageDirectory(),
photoName.getText().toString()+".jpg");
                    FileOutputStream outStream=null;
                    try
                    {  //打开指定文件对应的输出流
                        outStream=new FileOutputStream(file);
                        //把位图输出到指定文件中
                        bm.compress(CompressFormat.JPEG, 100,outStream);
                        outStream.close();
                    }
                    catch (IOException e)
                    {
                        e.printStackTrace();
                    }
                }
            }).setNegativeButton("取消", null).show();
        camera.stopPreview();          //重新浏览
        camera.startPreview();
        isPreview=true;
    }
};
}
```

该程序的运行效果如图 5-9 所示。

图 5-9　照相机案例布局主界面

5.2.7　控制摄像头录制视频短片

MediaRecorder 除了可用于录制音频之外，还可用于录制视频。使用 MediaRecorder 录制视频与录制音频的步骤基本相同。只是录制视频时不仅需要采集声音，还需要采集图像。为了让 MediaRecorder 录制时采集图像，应该在调用 setAudioSource (int audio_source)方法时再调用 setVideoSource(int video_source) 方法来设置图像来源。除此之外，还需要调用 setOutputFormat() 方法设置输出文件格式，之后进行如下步骤：

（1）调用 MediaRecorder 对象的 setVideoEncoder()、setVideoEncodingBitRate(int bitRate)、setVideoFrameRate 设置所录制的视频的编码格式、编码位率、每秒多少帧等，这些参数将可以控制所录制的视频的品质、文件的大小。一般来说，视频品质越好，视频文件越大。

（2）调用 MediaRecorder 对象的 setPreviewDisplay(Surface sv)方法设置使用哪个 SurfaceView 显示视频预览。

（3）之后的步骤基本与录制音频相同。

MediaRecorder 典型应用案例：迷你录像机

该案例实现视频的录制，界面中提供了两个按钮用于控制开始、结束录制。该程序的界面布局图如图 5-10 所示，其详细代码请参见配套光盘中的 CH05_7。

图 5-10　录制视频短片案例布局主界面

程序的界面布局代码如下：CH05\CH05_7\res\layout\ activity_main.xml。

```
<?xml version="1.0" encoding="utf-8"?>
<RelativeLayout xmlns:android="http://schemas.android.com/apk/res/android"
  android:orientation="vertical"
  android:layout_width="match_parent"
  android:layout_height="match_parent">
<!-- 显示视频预览的 SurfaceView -->
<SurfaceView
  android:id="@+id/sView"
  android:layout_width="match_parent"
  android:layout_height="match_parent" />
<LinearLayout
  android:orientation="horizontal"
  android:layout_width="wrap_content"
  android:layout_height="wrap_content"
  android:gravity="center_horizontal"
  android:layout_alignParentBottom="true"
  android:layout_centerHorizontal="true">
  <ImageButton
    android:id="@+id/record"
    android:layout_width="wrap_content"
```

```
        android:layout_height="wrap_content"
        android:src="@drawable/record" />
    <ImageButton
        android:id="@+id/stop"
        android:layout_width="wrap_content"
        android:layout_height="wrap_content"
        android:src="@drawable/stop" />
</LinearLayout>
</RelativeLayout>
```

程序的 Java 代码如下：CH05 \CH05_7\src\com\zigcloud\sound\ RecordVideo.java。

```
public class RecordVideo extends Activity implements OnClickListener
{
    ImageButton record , stop;                    //程序中的两个按钮
    File videoFile ;                              //系统的视频文件
    MediaRecorder mRecorder;
    SurfaceView sView;                            //显示视频预览的 SurfaceView
    private boolean isRecording=false;    //记录是否正在进行录制
    public void onCreate(Bundle savedInstanceState)
    {
        super.onCreate(savedInstanceState);
        setContentView(R.layout.activity_main);
        //获取程序界面中的两个按钮
        record=(ImageButton) findViewById(R.id.record);
        stop=(ImageButton) findViewById(R.id.stop);
        stop.setEnabled(false);                    //让 stop 按钮不可用
        //为两个按钮的单击事件绑定监听器
        record.setOnClickListener(this);
        stop.setOnClickListener(this);
        sView=(SurfaceView) this.findViewById(R.id.sView);
                                                  //获取程序界面中的 SurfaceView
        //设置 Surface 不需要自己维护缓冲区
        sView.getHolder().setType(SurfaceHolder.SURFACE_TYPE_PUSH_BUFFERS);
        sView.getHolder().setFixedSize(320, 280);       //设置分辨率
        sView.getHolder().setKeepScreenOn(true); //设置该组件让屏幕不会自动关闭
    }
    public void onClick(View source)
    {
        switch (source.getId())
        {
          case R.id.record:                    //单击录制按钮
            if (!Environment.getExternalStorageState().equals(android.os.
Environment.MEDIA_MOUNTED))
            {
                Toast.makeText(RecordVideo.this, "SD 卡不存在，请插入 SD 卡！",
Toast.LENGTH_SHORT).show();
                return;
            }
            try
            {  //创建保存录制视频的视频文件
                videoFile=new File(Environment  .getExternalStorageDirectory()
```

```
.getCanonicalFile() + "/myvideo.mp4");
                mRecorder=new MediaRecorder();  //创建 MediaPlayer 对象
                mRecorder.reset();
                //设置从麦克风采集声音
                mRecorder.setAudioSource(MediaRecorder.AudioSource.MIC);
                //设置从摄像头采集图像
                mRecorder.setVideoSource(MediaRecorder.VideoSource.CAMERA);
                //设置视频文件的输出格式
                //必须在设置声音编码格式、图像编码格式之前设置
                mRecorder.setOutputFormat(MediaRecorder.OutputFormat.MPEG_4);
                //设置声音编码的格式

                mRecorder.setAudioEncoder(MediaRecorder.AudioEncoder.DEFAULT);
                //设置图像编码的格式

                mRecorder.setVideoEncoder(MediaRecorder.VideoEncoder.MPEG_4_SP);
                mRecorder.setVideoSize(320, 280);
                mRecorder.setVideoFrameRate(4);    // 每秒 4 帧
                mRecorder.setOutputFile(videoFile.getAbsolutePath());
                //指定使用 SurfaceView 来预览视频

                mRecorder.setPreviewDisplay(sView.getHolder().getSurface());
                mRecorder.prepare();
                mRecorder.start();                    //开始录制
                System.out.println("---正在录制---");
                record.setEnabled(false);             //让 record 按钮不可用
                stop.setEnabled(true);                //让 stop 按钮可用
                isRecording=true;
            }
            catch (Exception e)
            {
                e.printStackTrace();
            }
            break;
        case R.id.stop:          //单击停止按钮
            if (isRecording)     //如果正在进行录制
            {
                mRecorder.stop();        //停止录制
                mRecorder.release();     //释放资源
                mRecorder=null;
                record.setEnabled(true);
                            //让 record 按钮可用
                stop.setEnabled(false);
                            //让 stop 按钮不可用
            }
            break;
        }
    }
}
```

该程序的运行效果如图 5-11 所示。

图 5-11　录制视频短片案例运行主界面

5.3 案例——流媒体播放器

在开发一些视频监控软件客户端、网络电视客户端时，要在 Android 客户端实时显示摄像头或者视频资源的画面，就会涉及多媒体部分的知识，Android 中也提供了对流媒体播放的支持，下面的实例将在 Android 中实现播放流媒体资源（RTSP 的地址等，现在很多摄像头都提供 RTSP 格式的输出，很多网络电视台也会以 RTSP 格式发布视频）。该程序的界面布局图如图 5-12 所示，其详细代码请参见配套光盘中的 CH05_8。

图 5-12　流媒体播放案例界面

程序的界面布局代码如下：CH05\ CH05_8\res\layout\ activity_main.xml。

```
<RelativeLayout xmlns:android="http://schemas.android.com/apk/res/android"
    xmlns:tools="http://schemas.android.com/tools"
    android:layout_width="match_parent"
    android:layout_height="match_parent"
    android:paddingBottom="@dimen/activity_vertical_margin"
    android:paddingLeft="@dimen/activity_horizontal_margin"
    android:paddingRight="@dimen/activity_horizontal_margin"
    android:paddingTop="@dimen/activity_vertical_margin"
    tools:context=".MainActivity" >
    <LinearLayout
        android:layout_width="fill_parent"
        android:layout_height="wrap_content"
        android:layout_alignParentTop="true"
        android:layout_centerHorizontal="true"
        android:gravity="center_vertical"  >
        <Button
        android:id="@+id/btnPlayUrl"
        android:layout_width="wrap_content"
        android:layout_height="wrap_content"
        android:text="@string/play" >
    </Button>
    <SeekBar
        android:id="@+id/skbProgress"
        android:layout_width="fill_parent"
        android:layout_height="wrap_content"
        android:max="100"
        android:paddingLeft="10dip"
        android:paddingRight="10dip" />
    </LinearLayout>
</RelativeLayout>
```

程序的 Java 代码：CH05\CH05_8\src\com\example\demovideostream\MainActivity.java。

```
public class MainActivity extends Activity implements OnClickListener {
    /***流媒体播放实例* */
    private Button btnPlayUrl;
    private SeekBar skbProgress;
    private Player player;
```

```
    private int progress;
    protected void onCreate(Bundle savedInstanceState) {
        super.onCreate(savedInstanceState);
        setContentView(R.layout.activity_main);
        btnPlayUrl=(Button) this.findViewById(R.id.btnPlayUrl);
        btnPlayUrl.setOnClickListener(this);
        skbProgress=(SeekBar) this.findViewById(R.id.skbProgress);
        skbProgress.setOnSeekBarChangeListener(new SeekBarChangeEvent());
        player=new Player(skbProgress);
    }
    public boolean onCreateOptionsMenu(Menu menu) {
        getMenuInflater().inflate(R.menu.main, menu);
                                        //膨胀的菜单，将菜单项添加到行动条
        return true;
    }
    public void onClick(View arg0) {
        int id=arg0.getId();
        switch (id) {
            case R.id.btnPlayUrl:
                //网络音频地址
                String url="rtsp://211.139.194.251:554/live/2/13E6330A3119 3128/
5iLd2iNl5nQ2s8r8.sdp";
                player.playUrl(url);
                break;
            default:
                break;
        }
    }
    /*** SeekBar 事件监听类* */
    private class SeekBarChangeEvent implements SeekBar.OnSeekBar ChangeListener {
        public void onProgressChanged(SeekBar seekBar,int progress,boolean
fromUser) {
            MainActivity.this.progress=progress * player.mediaPlayer.
getDuration() / seekBar.getMax();
        }
        public void onStartTrackingTouch(SeekBar seekBar) {
        }
        public void onStopTrackingTouch(SeekBar seekBar) {
            player.mediaPlayer.seekTo(progress);
        }
    }
}
```

程序清单：CH05\ CH05_8\com\example\demovideostream\mediaPlayer.java。

```
public class Player implements MediaPlayer.OnPreparedListener{
    /***自定义播放器* */
    public MediaPlayer mediaPlayer;
```

```
    private SeekBar skbProgress;
    private Timer mTimer=new Timer();
    public Player(SeekBar skbProgress)
    {
        this.skbProgress=skbProgress;
        try {
            mediaPlayer=new MediaPlayer();
            mediaPlayer.setAudioStreamType(AudioManager.STREAM_MUSIC);
            mediaPlayer.setOnPreparedListener(this);
        } catch (Exception e) {
            Log.e("mediaPlayer", "error", e);
        }
        mTimer.schedule(mTimerTask, 0, 10);
    }
    TimerTask mTimerTask=new TimerTask() {
        public void run() {
            if(mediaPlayer==null)
                return;
            if(mediaPlayer!=null){
                if(mediaPlayer.isPlaying() && skbProgress.isPressed()== false) {
                    handleProgress.sendEmptyMessage(0);
                }
            }
        }
    };
    Handler handleProgress=new Handler() {
        public void handleMessage(Message msg) {
            if(mediaPlayer==null){
                return;
            }
            int position=mediaPlayer.getCurrentPosition();
            int duration=mediaPlayer.getDuration();
            if(duration>0) {
                long pos=skbProgress.getMax() * position / duration;
                skbProgress.setProgress((int) pos);
            }
        };
    };
    public void playUrl(String videoUrl) /*** 根据传入的网址播放视频* */
    {
        try {
            mediaPlayer.reset();
            mediaPlayer.setDataSource(videoUrl);
            mediaPlayer.prepareAsync();
        } catch (IllegalArgumentException e) {
            e.printStackTrace();
```

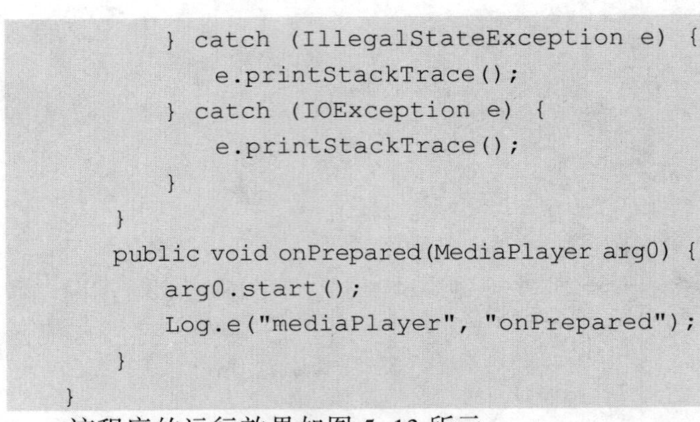

```
        } catch (IllegalStateException e) {
            e.printStackTrace();
        } catch (IOException e) {
            e.printStackTrace();
        }
    }
    public void onPrepared(MediaPlayer arg0) {
        arg0.start();
        Log.e("mediaPlayer", "onPrepared");
    }
}
```

该程序的运行效果如图 5-13 所示。

图 5-13　流媒体播放案例运行主界面

5.4 知识扩展

5.4.1 传感器知识

（1）加速度传感器（SENSOR_TYPE_ACCELEROMETER）。

（2）磁力传感器（SENSOR_TYPE_MAGNETIC_FIELD）。

（3）方向传感器（SENSOR_TYPE_ORIENTATION）。

（4）陀螺仪传感器（SENSOR_TYPE_GYROSCOPE）。

5.4.2 传感器的典型案例

（1）智能灯。

（2）老人防护仪。

5.5 本章小结

音频和视频都是非常重要的多媒体形式，Android 系统为音频、视频等多媒体的播放、录制提供了强大的支持。学习本章需要重点掌握如何使用 MediaPlayer、SoundPool 播放音频，如何使用 VideoView、MediaPlayer 播放视频。除此之外，还需要掌握通过 MediaRecorder 录制音频的方法，以及控制摄像头拍照、录制视频的方法。

5.6 强化练习

1. 填空题

（1）MediaPlayer 的三个主要的方法是（　　　）、（　　　）和（　　　）。

（2）MediaPlayer 存在缺点有（　　　）和（　　　）。

（3）（　　　）主要用于播放一些较短的声音片段，与 MediaPlayer 相比，它的优势在于 CPU 资源占用量低和反应延迟小。

（4）为了在 Android 应用中播放视频，Android 提供了（　　　）组件，它就是一个位于 android.widget 包下的组件，它的作用与 ImageView 类似，只是 ImageView 用于显示图片，而它用于播放视频。

（5）使用 VideoView 播放视频简单、方便，但是早期开发者更喜欢使用 MediaPlayer 播放视频。但由于 MediaPlayer 主要用于播放音频，因此它没有提供图像输出界面，此时就需要借助于（　　）来显示 MediaPlayer 播放的图像输出。

（6）手持设备一般都提供了麦克风硬件，而 Android 系统可以利用硬件来录制音频。为了在 Android 系统中录制音频，Android 提供了（　　）类。

2. 问答题

（1）列举出 Android 中播放音频相关的类。

（2）列举出 Android 中播放视频相关的类。

（3）列举出 Android 中摄像头相关的类。

第6章 Android 系统服务应用

学习目标：

- 掌握活动管理器（ActivityManager）的使用；
- 掌握警报管理器（AlarmManager）的使用；
- 掌握音频管理器（AudioManager）的使用；
- 掌握剪贴板管理器（ClipboardManager）的使用；
- 掌握通知管理器（NotificationManager）的使用。

6.1 学习导入

Android 系统服务与前面所讲过的服务是不同的，系统服务不仅指服务组件，而且还包括 Android 系统提供的服务功能。它们的使用方式需要系统提供的特定方式来获取希望得到的服务功能接口，通过这些接口与系统的核心组件进行交互。

通过上述系统服务接口可以方便地获取系统信息，对系统功能进行集成。从 Android SDK1.5 以后，Android 系统的服务接口都由 Context 类提供。通过 getSystemService() 方法，开发者就可以通过指定的服务字符串标识获取相应的服务。下面通过表 6-1 了解在本章的学习中将会用到的服务标识符。

表 6-1　Android 系统服务

服务字符串标识	说　明	服务字符串标识	说　明
ACTIVITY_SERVICE	管理 Activity 的服务	NOTIFICATION_SERVICE	通知服务
ALARM_SERVICE	闹钟（警报）服务	POWER_SERVICE	电源管理服务
AUDIO_SERVICE	音量控制服务	SEARCH_SERVICE	搜索服务
CLIPBOARD_SERVICE	剪贴板服务	SENSOR_SERVICE	传感器服务
CONNECTIVITY_SERVICE	连接管理服务	TELTPHONY_SERVICE	电话信息服务
INPUT_METHOD_SERVICE	输入法服务	VIBRATOR_SERVICE	振动器服务
KEYGUARD_SERVICE	键盘锁定服务	WALLPAPER_SERVICE	墙纸服务
LAYOUT_INFLATER_SERVICE	布局填充服务	WIFI_SERVICE	Wi-Fi 服务
LOCATION_SERVICE	定位服务	WINDOW_SERVICE	窗口服务

表 6-1 中所列出来的就是要经常用到的系统服务的标识。下面详细讲解如何使用这些服务。

6.2 技术准备

6.2.1 活动管理器（ActivityManager）

ActivityManager 是对所有运行中的 Activity 组件进行管理，在 android.app 包中。通过 ActivityManager 可以获取当前设备的配置信息、内存信息、进程错误状态、近期任务、运行中进程、运行中服务和运行中任务信息。

通过 getSystemService()方法获取系统服务——ACTIVITY_SERVICE；在得到系统服务后，调用 getDeviceConfigurationInfo()方法获取配置信息接口，在配置信息中包括系统中的多项配置信息，包括输入方式类型、键盘类型、导航方式类型、触摸屏方式类型等内容。

（1）输入方式类型。通过设置配置信息接口的 reqInputFeatures 属性可以获得当前设备的输入方式，其输入方式的类型在 ConfigurationInfo 接口中定义。一般有以下两种：

- INPUT_FEATURE_FIVE_WAY_NAV　五向导航键输入。
- INPUT_FEATURE_HARD_KEYBOARD　硬键盘输入。

（2）键盘类型。通过配置信息接口的 reqKeyboardType 属性可以获得当前设备的键盘类型，键盘类型主要有以下几种：

- KEYBOARD_UNDEFINED：未定义键盘。
- KEYBOARD_NOKEYS：无键键盘。
- KEYBOARD_QWERTY：打字机键盘。
- KEYBOARD_12KEY：十二键盘。

（3）导航方式类型。同样通过配置信息接口的 reqNavigation 属性可以获得当前设备的导航方式，有以下几种：

- NAVIGATION_UNDEFINED：未定义导航。
- NAVIGATION_DPAD：面板导航。
- NAVIGATION_TRACKBALL：定位球导航。
- NAVIGATION_WHEEL：滚轮导航。

（4）触摸屏方式类型。触摸屏类型的属性为 reqTouchScreen。主要分为：

- TOUCHSCREEN_NOTOUCH：不支持触摸屏。
- TOUCHSCREEN_STYLUS：触摸笔。
- TOUCHSCREEN_FINGER：手指触摸。

除了调用配置信息接口获取系统的配置信息外，还可以通过 ActivityManager 中的 getMemoryInfo()方法获取系统的内存信息接口。如下代码所示：

```
MemoryInfo memInfo=new MemoryInfo();
service.getMemoryInfo(memInfo);
```

当获取到内存信息接口以后，就可以获取当前的内存信息，其接口的属性如下：

- availMen：可用内存。
- lowMemory：是否低内存。
- threshold：内存阈值。

除此之外，同样可以通过 getProgressesInErrorState()方法获取系统进程的错误状态信息

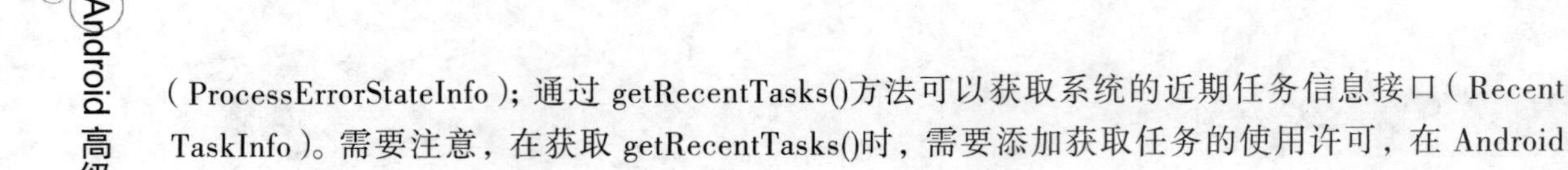

(ProcessErrorStateInfo)；通过 getRecentTasks()方法可以获取系统的近期任务信息接口（RecentTaskInfo）。需要注意，在获取 getRecentTasks()时，需要添加获取任务的使用许可，在 AndroidManifest.xml 文件中添加：

```
<uses-permission android:nam="android.permission.GET_TASKS">
```

ActivityManager 典型应用案例：手持设备活动管理器

程序实例 ActivityManagerTest 为对 ActivityManager 应用的一个简单实例，它在主程序中如何调用所需要信息，其代码如下：

```
public class ActivityManagerTest extends Activity {
   TextView txtView=null;
   String strTxt=null;
    public void onCreate(Bundle savedInstanceState) {
        /** 第一次创建活动（activity）时调用*/
        super.onCreate(savedInstanceState);
        setContentView(R.layout.main);
        txtView=(TextView)findViewById(R.id.txtView);
        ActivityManager service=(ActivityManager)getSystemService(Context.ACTIVITY_SERVICE);
        ConfigurationInfo cfgInfo=service.getDeviceConfigurationInfo();
        if(cfgInfo.reqInputFeatures==ConfigurationInfo.INPUT_FEATURE_HARD_KEYBOARD){
            strTxt="硬键盘输入";
        }if(cfgInfo.reqInputFeatures==ConfigurationInfo.INPUT_FEATURE_FIVE_WAY_NAV){
            strTxt="五向导航键输入";
        }
        if(cfgInfo.reqKeyboardType==Configuration.KEYBOARD_UNDEFINED){
            strTxt+=",未定义键盘";
        }if(cfgInfo.reqKeyboardType==Configuration.KEYBOARD_NOKEYS){
            strTxt+=", 无键键盘";
        }
        MemoryInfo memInfo=new MemoryInfo();
        service.getMemoryInfo(memInfo);
        System.out.println("总内存量: "+memInfo.availMem);
        System.out.println("是否低内存"+memInfo.lowMemory);
        System.out.println("是否阈值"+memInfo.threshold);
    }
}
```


6.2.2 警报管理器（AlarmManager）

AlarmManager 用于访问系统的警报服务，同样位于 android.app 包中。警报管理器允许用户预定自定义应用程序的运行时间。为了获取 AlarmManager，同样需要调用 getSystemService()方法，只是需要的参数为 Context.ALARM_SERVICE。在获取到 AlarmManager 对象后，需要设置警报的时区，以便能准确计时。需调用其 setTimeZone()方法。

设置警报主要分为两种：一次性警报和周期性警报。一次性警报，顾名思义该提示框只会出现一次。通过警报管理器接口的 set()方法确定一个警报时间。该方法包含 3 个参数：第一个参数是时间标志，用以确定警报的计时和报警方式。其参数说明如下：

- ELAPSED_REALTIME：从系统启动开始计时（包括休眠时间）。
- ELAPSED_REALTIME_WAKEUP：从系统启动开始计时（包括休眠时间）并唤醒系统。
- RTC：以系统当前的时间戳计时（UTC 格式）。
- RTC_WAKEUP：以系统当前的时间戳计时（UTC 格式）并唤醒系统。

第二个参数是警报触发的时间点；第三个参数是一个未决意向，用于指明警报的处理方式。这里需要说明一下：讲到第三个参数是一个未决意向，也就是说这个行为并不一定是一定要发生的。例如，定义了一个闹钟是过 10 min 后触发的，但是这个行为可能在过了 5 min 后就取消了。但是，在定义时同样需要确定行为的执行者。设置方式如下所示：

```
PendingIntent op=new PendingIntent();
service.set(AlarmManager.RTC_WAKEUP, System.currentTimeMillis(), op);
```

周期性警报与一次性警报的区别就在于该提示框会按照指定的时间间隔出现。周期性警报是通过警报管理器的 setRepeating()方法设置，该方法包含 4 个参数，与前面的设置相比多了一个警报时间的触发间隔时间参数。代码如下所示：

```
setRepeating(AlarmManager.RTC_WAKEUP,System.currentTimeMillis()+(5*1000)),
(5*1000),mPendingIntent);
```

上面介绍了设置警报的知识。当然，也可以取消警报设置。相比设置，取消警报比较简单，通过调用 AlarmManager.cancel(Intent intent)即可。需要注意在 AndroidManifest.xml 添加：

```
<uses-permission android:name="android.permission.SET_TIME_ZONE"/>
```

AlarmManager 典型应用案例：手持设备警报管理器

该案例利用 AlarmManager 控件实现手持设备的警报管理器，程序运行界面如图 6-1 所示。

图 6-1　程序运行图

在程序中通过重写一个 BroadcastReceiver 来对广播事件进行监听。首先注册广播接收器，从而让程序知道有自定义的广播接收器组件。然后就是设置主程序。在主程序中就用到了 AlarmManager。主程序的实现代码如下：CH06\CH06_1\src\sziit\ex\ch06\CH06_1.java。

```
public class CH06_1 extends Activity {
  public Button btnOnce;
  public Button btnRepeat;
  public Button btnCancel;
  private AlarmManager mManager=null;
  private PendingIntent mPendingIntent=null;
  public void onCreate(Bundle savedInstanceState) {
  /***第一次创建活动（activity ）时调用**/
      super.onCreate(savedInstanceState);
      setContentView(R.layout.main);
      mManager=(AlarmManager)getSystemService(Context.ALARM_SERVICE);
      mManager.setTimeZone("GMT+08:00");
      btnOnce=(Button)findViewById(R.id.Once);
      btnRepeat=(Button)findViewById(R.id.Repeat);
      btnCancel=(Button)findViewById(R.id.Cancel);
      Intent intent=new Intent(this, AlarmListener.class);
      mPendingIntent=PendingIntent.getBroadcast(this, 0, intent, 0);
      btnOnce.setOnClickListener(new Button.OnClickListener(){
```

```
            public void onClick(View v) {
                doOnce();
            }
        });
        btnRepeat.setOnClickListener(new Button.OnClickListener(){
            public void onClick(View v) {
                doRepeat();
            }
        });
        btnCancel.setOnClickListener(new Button.OnClickListener(){
            public void onClick(View v) {
                doCancel();
            }
        });
    }
    protected void doCancel() {
        mManager.cancel(mPendingIntent);
    }
    protected void doRepeat() {
        mManager.setRepeating(AlarmManager.RTC_WAKEUP,System.currentTimeMillis()
+ (5*1000), (5*1000), mPendingIntent);          //间隔
    }
    protected void doOnce() {
        mManager.set(AlarmManager.RTC_WAKEUP, System.currentTimeMillis()
+ (5*1000), mPendingIntent);    //触发事件（5 秒之后）
    }
}
```

另外，定义的广播接收器代码如下：CH06\CH06_1\src\sziit\ex\ch06\AlarmListener.java。

```
public class AlarmListener extends BroadcastReceiver {
  public void onReceive(Context context, Intent intent) {
     Toast.makeText(context, "Ding-ding", Toast.LENGTH_LONG).show();
  }
}
```

最后，在 AndroidManifest.xml 中声明广播接收器组件即可：

```
<receiver android:label="AlarmListener" android:name=".AlarmListener"/>
```

AlarmManager 应用案例：手持设备壁纸定时切换

该程序在界面中放置两个按钮，一个按钮用于启动定时更换壁纸，另一个按钮用于关闭定时更换壁纸。该程序的界面布局如图 6-2 所示，其详细代码请参见配套光盘中的 CH06_2。

图 6-2　定时更换壁纸案例运行主界面

程序的界面布局代码如下：CH06\CH06_2\res\layout\activity_main.xml。

```
<?xml version="1.0" encoding="utf-8"?>
<LinearLayout xmlns:android="http://schemas.android.com/apk/res/android"
  android:orientation="horizontal"
  android:layout_width="fill_parent"
  android:layout_height="fill_parent"
  android:gravity="center">
```

```
<Button
  android:id="@+id/start"
  android:layout_width="wrap_content"
  android:layout_height="wrap_content"
  android:text="@string/start"/>
<Button
  android:id="@+id/stop"
  android:layout_width="wrap_content"
  android:layout_height="wrap_content"
  android:text="@string/stop"/>
</LinearLayout>
```

程序的 Java 代码如下：CH06\ CH06_2\src\com\zigcloud\manager\ AlarmChangeWallpaper.java。

```
public class AlarmChangeWallpaper extends Activity
{ //定义 AlarmManager 对象
    AlarmManager aManager;
    Button start, stop;
    public void onCreate(Bundle savedInstanceState)
    {
     super.onCreate(savedInstanceState);
     setContentView(R.layout.activity_main);
     start=(Button) findViewById(R.id.start);
     stop=(Button) findViewById(R.id.stop);
     aManager=(AlarmManager) getSystemService(Service.ALARM_SERVICE);
     //指定启动 ChangeService 组件
     Intent intent=new Intent(AlarmChangeWallpaper.this,ChangeService. class);
     //创建 PendingIntent 对象
     final PendingIntent pi=PendingIntent.getService(AlarmChangeWallpaper.
this, 0, intent, 0);
     start.setOnClickListener(new OnClickListener()
     {
        public void onClick(View arg0)
        {  //设置每隔 5 秒执行 pi 代表的组件一次
           aManager.setRepeating(AlarmManager.RTC_WAKEUP, 0, 5000, pi);
           start.setEnabled(false);
           stop.setEnabled(true);
           Toast.makeText(AlarmChangeWallpaper.this, "壁纸定时更换启动成功啦",
Toast.LENGTH_SHORT).show();
        }
     });
     stop.setOnClickListener(new OnClickListener()
     {
        public void onClick(View arg0)
        {
           start.setEnabled(true);
           stop.setEnabled(false);
           aManager.cancel(pi);       //取消对 pi 的调度
        }
     });
  }
}
```

程序清单：CH06\ CH06_2\src\com\zigcloud\manager\ ChangeService.java。

```
public class ChangeService extends Service
{   //定义定时更换的壁纸资源
    int[] wallpapers=new int[]{
       R.drawable.shuangta,
       R.drawable.lijiang,
       R.drawable.qiao,
       R.drawable.shui
    };
    WallpaperManager wManager;              //定义系统的壁纸管理服务
    int current=0;                          //定义当前所显示的壁纸
    public int onStartCommand(Intent intent, int flags, int startId)
    {//如果到了最后一张，系统重新开始
     if(current>=4)
        current=0;
     try
     {
        wManager.setResource(wallpapers[current++]);     //改变壁纸
     }
     catch(Exception e)
     {
        e.printStackTrace();
     }
     return START_STICKY;
    }
    public void onCreate()
    {
     super.onCreate();
     wManager=WallpaperManager.getInstance(this);
                                   // 初始化 WallpaperManager
  }
  public IBinder onBind(Intent intent)
  {
     return null;
  }
}
```

6.2.3 音频管理器（AudioManager）

AudioManager（音频管理器）提供了访问音量和响铃模式的控制，定义在 android.media 包中。可以通过 AudioManager 获取和设置音频以及音量。跟前面其他管理器的使用方法相似，通过 getSystemService()方法获取音频管理器接口。然后就可以调用 AudioManager 对象的方法实现对音量的设置和管理。

（1）获取音量设置。获取音量设置包括两种类型，获取系统最大音量和获取当前音量值。获取音量的最大值可以通过音量管理器接口的 getStreamMaxVolume()方法获得指定音频流的音量最大值。该方法仅有的一个参数就是音频流类型，如下：

- STREAM_VOICE_CALL：呼叫声音。
- STREAM_SYSTEM：系统声音。
- STREAM_RING：用于响铃的音频流。
- STREAM_MUSIC：用于音乐的音频流。
- STREAM_ALARM：用于警报的音频流。

（2）获取当前音量。通过 getStreamVolume()方法可以获得指定音频流的当前音量设定值。其音频类型上面已经介绍过。

（3）调整音量设置。通过音量管理器接口的 adjustStreamVolume()方法可以调整指定音频流的当前音量设定值。该方法有三个参数：第一个参数是音频流的类型；第二个参数是音量调节的方向；第三个参数是控制标志。第一个参数音频流不再赘述，第二个参数是音量调节的方向，其中音量调节方向的定义如下：

- ADJUST_LOWER：调低。
- ADJUST_RAISE：调高。
- ADJUST_SAME：不调整。

关于音量控制标志的参数如下：

- FLAG_ALLOW_RINGER_MODES：是否包含响铃模式选项。
- FLAG_PLAY_SOUND：当改变音量的时候是否播放声音。
- FLAG_REMOVE_SOUND_AND_VIBRATE：是否移除队列中的任何声音或震动。
- FLAG_SHOW_UI：是否显示音量调节滑动条。
- FLAG_VIBRATE：是否进入震动响铃模式。

AudioManager 的典型应用案例：音频管理器

该案例利用 AudioManager 控件实现典型的移动设备的音频管理器，程序运行结果如图 6-3 所示。

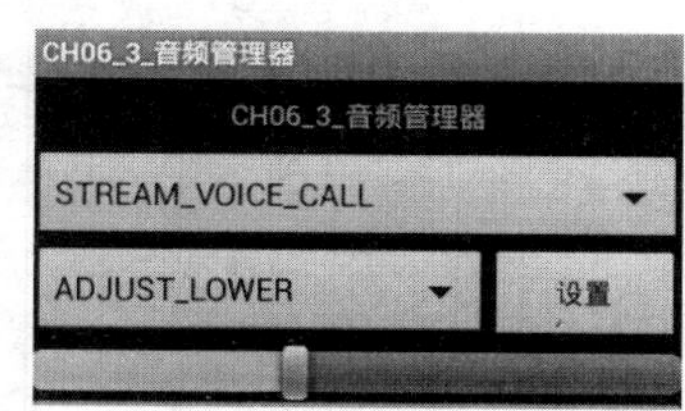

图 6-3　程序运行图

在这个程序中用了几个下拉列表，程序的界面布局代码如下：CH06\CH06_3\res\layout\main.xml。

```
<?xml version="1.0" encoding="utf-8"?>
<LinearLayout xmlns:android="http://schemas.android.com/apk/res/android"
    android:orientation="vertical"
    android:layout_width="fill_parent"
    android:layout_height="fill_parent" >
  <TextView
      android:layout_width="fill_parent"
      android:layout_height="wrap_content"
      android:text="@string/app_name"
      android:padding="8sp"
      android:gravity="center_horizontal" />
  <Spinner android:id="@+id/SPN_ITEMS"
     android:layout_width="fill_parent"
     android:layout_height="wrap_content"
     android:entries="@array/audio_streams" />
   <TableLayout
```

```
    android:layout_width="fill_parent"
    android:layout_height="wrap_content"
    android:stretchColumns="0,1" >
    <TableRow>
        <Spinner android:id="@+id/SPN_ITEMS2"
            android:layout_width="fill_parent"
            android:layout_height="wrap_content"
            android:entries="@array/directions" />
        <Button android:id="@+id/BTN_SET"
            android:layout_width="fill_parent"
            android:layout_height="wrap_content"
            android:text="设置" />
    </TableRow>
  </TableLayout>
  <SeekBar android:id="@+id/BAR_VOLUME"
    android:layout_width="fill_parent"
    android:layout_height="wrap_content"
    android:max="100"
    android:stepSize="10"
    android:progress="30" />
  <EditText android:id="@+id/TXT_CONTENTS"
    android:layout_width="fill_parent"
    android:layout_height="fill_parent"
    android:scrollbars="vertical"
    android:editable="false"
    android:textSize="6pt" />
</LinearLayout>
```


在主程序中应用 AudioManager。代码如下：CH06\CH06_3\src\sziit\ex\ ch06\CH06_3.java。

```
public class CH06_3 extends Activity implements OnClickListener, OnItem
SelectedListener {
    private EditText mTxtContents=null;
    private Spinner mItems=null;
    private Spinner mDirections=null;
    private Button mBtnSet=null;
    private SeekBar mBarVolume=null;
    private AudioManager mService=null;
    public void onCreate(Bundle savedInstanceState) {
    /**第一次创建活动（activity）时调用*/
        super.onCreate(savedInstanceState);
        setContentView(R.layout.main);
        mTxtContents=(EditText)findViewById(R.id.TXT_CONTENTS);
        mItems=(Spinner)findViewById(R.id.SPN_ITEMS);
        mDirections=(Spinner)findViewById(R.id.SPN_ITEMS2);
        mBtnSet=(Button)findViewById(R.id.BTN_SET);
        mBarVolume=(SeekBar)findViewById(R.id.BAR_VOLUME);
        mBtnSet.setOnClickListener(this);
```

```
        mItems.setOnItemSelectedListener(this);
        //获取音频服务管理器
        mService=(AudioManager)(this.getSystemService(Context.AUDIO_SERVICE) );
        mItems.setSelection(0);                          //初始化
        mDirections.setSelection(0);
    }
    //当流类型选择改变时
    public void onItemSelected(AdapterView<?> parent,View v,int pos,long id){
        int selectedPos=mItems.getSelectedItemPosition();
        String[] items=getResources().getStringArray(R.array.audio_streams);
        final String streamType=items[selectedPos];
        if(streamType.equalsIgnoreCase("STREAM_VOICE_CALL")) {
            int max =mService.getStreamMaxVolume(AudioManager.STREAM_VOICE_CALL);
            mBarVolume.setMax(max);
            mBarVolume.setProgress(max);
        }
        else if(streamType.equalsIgnoreCase("STREAM_SYSTEM")) {
            int max= mService.getStreamMaxVolume(AudioManager.STREAM_SYSTEM);
            mBarVolume.setMax(max);
            mBarVolume.setProgress(max);
        }
        else if(streamType.equalsIgnoreCase("STREAM_RING")) {
            int max=mService.getStreamMaxVolume(AudioManager.STREAM_RING);
        mBarVolume.setMax(max);
            mBarVolume.setProgress(max);
        }
        else if(streamType.equalsIgnoreCase("STREAM_MUSIC")) {
            int max=mService.getStreamMaxVolume(AudioManager.STREAM_MUSIC);
            mBarVolume.setMax(max);
            mBarVolume.setProgress(max);
        }
        else if(streamType.equalsIgnoreCase("STREAM_ALARM")) {
            int max=mService.getStreamMaxVolume(AudioManager.STREAM_ALARM);
            mBarVolume.setMax(max);
            mBarVolume.setProgress(max);
        }
        getSetting(items[selectedPos]);
    }
    private void getSetting(String streamType) {           //获取设置
        if(streamType.equalsIgnoreCase("STREAM_VOICE_CALL")) {
            int volume =mService.getStreamVolume(AudioManager.STREAM_VOICE_CALL);
            mBarVolume.setProgress(volume);
            clearText();
            printText("STREAM_VOICE_CALL volume: "+volume+"/"+mBarVolume. getMax());
        }
        else if(streamType.equalsIgnoreCase("STREAM_SYSTEM")) {
```

```
        int volume=mService.getStreamVolume(AudioManager.STREAM_SYSTEM);
        mBarVolume.setProgress(volume);
        clearText();
        printText("STREAM_SYSTEM volume: "+volume+"/"+mBarVolume. getMax());
    }
    else if(streamType.equalsIgnoreCase("STREAM_RING")) {
        int volume=mService.getStreamVolume(AudioManager.STREAM_RING);
    mBarVolume.setProgress(volume);
        clearText();
        printText("STREAM_RING volume: "+volume+"/"+mBarVolume. getMax());
    }
    else if(streamType.equalsIgnoreCase("STREAM_MUSIC")) {
        int volume=mService.getStreamVolume(AudioManager.STREAM_MUSIC);
        mBarVolume.setProgress(volume);
        clearText();
        printText("STREAM_MUSIC volume: "+volume+"/"+mBarVolume. getMax());
    }
    else if(streamType.equalsIgnoreCase("STREAM_ALARM")) {
        int volume=mService.getStreamVolume(AudioManager.STREAM_ALARM);
        mBarVolume.setProgress(volume);
        clearText();
        printText("STREAM_ALARM volume: "+volume+"/"+mBarVolume. getMax());
    }
}
public void onNothingSelected(AdapterView<?> parent) {
}
public void onClick(View v) {
    switch(v.getId() ) {
        case R.id.BTN_SET: {
            doSet();
            break;
        }
    }
}
private void doSet() {                        //执行设置
    int selectedPos=mItems.getSelectedItemPosition();
    int selectedPos2=mDirections.getSelectedItemPosition();
    String[] items=getResources().getStringArray(R.array.audio_streams);
    String[] directions=getResources().getStringArray(R.array.directions);
    System.out.println("Item #"+selectedPos+",direction #"+ selectedPos2);
    setSetting(items[selectedPos], directions[selectedPos2]);
}
private void setSetting(String streamType, String direction) {
                                              //调整所选流类型的音量大小
    System.out.println("setSetting"+"("+streamType+","+direction+")");
    int direction2=0;
```

```
        if(direction.equalsIgnoreCase("ADJUST_LOWER") ) {//调整方向
            direction2=AudioManager.ADJUST_LOWER;
            System.out.print("Direction is ADJUST_LOWER");
        }
        else if(direction.equalsIgnoreCase("ADJUST_RAISE") ) {
            direction2=AudioManager.ADJUST_RAISE;
            System.out.print("Direction is ADJUST_RAISE");
        }
        else if(direction.equalsIgnoreCase("ADJUST_SAME") ) {
            direction2=AudioManager.ADJUST_SAME;
            System.out.print("Direction is ADJUST_SAME");
        }
        if(streamType.equalsIgnoreCase("STREAM_VOICE_CALL") ) {
            mService.adjustStreamVolume(AudioManager.STREAM_VOICE_CALL,
direction2, AudioManager.FLAG_REMOVE_SOUND_AND_VIBRATE);
            int volume=mService.getStreamVolume(AudioManager.STREAM_VOICE_CALL);
            mBarVolume.setProgress(volume);
            clearText();
            printText("STREAM_VOICE_CALL volume: "+volume+"/"+mBarVolume.getMax());
        }
        else if(streamType.equalsIgnoreCase("STREAM_SYSTEM") ) {
            mService.adjustStreamVolume(AudioManager.STREAM_SYSTEM, direction2,
AudioManager.FLAG_REMOVE_SOUND_AND_VIBRATE);
            int volume=mService.getStreamVolume(AudioManager.STREAM_SYSTEM);
            mBarVolume.setProgress(volume);
            clearText();
            printText("STREAM_SYSTEM volume: "+volume+"/"+mBarVolume.getMax());
        }
        else if(streamType.equalsIgnoreCase("STREAM_RING") ) {
            mService.adjustStreamVolume(AudioManager.STREAM_RING,direction2,
AudioManager.FLAG_REMOVE_SOUND_AND_VIBRATE);
            int volume=mService.getStreamVolume(AudioManager.STREAM_RING);
            mBarVolume.setProgress(volume);
            clearText();
            printText("STREAM_RING volume: "+volume+"/"+mBarVolume.getMax());
        }
        else if(streamType.equalsIgnoreCase("STREAM_MUSIC") ) {
            mService.adjustStreamVolume(AudioManager.STREAM_MUSIC, direction2,
AudioManager.FLAG_REMOVE_SOUND_AND_VIBRATE);
            int volume=mService.getStreamVolume(AudioManager.STREAM_MUSIC);
            mBarVolume.setProgress(volume);
            clearText();
            printText("STREAM_MUSIC volume: " + volume + "/" + mBarVolume.getMax() );
        }
        else if(streamType.equalsIgnoreCase("STREAM_ALARM") ) {
```

```
            mService.adjustStreamVolume(AudioManager.STREAM_ALARM, direction2,
AudioManager.FLAG_REMOVE_SOUND_AND_VIBRATE);
            int volume=mService.getStreamVolume(AudioManager.STREAM_ALARM);
            mBarVolume.setProgress(volume);
            clearText();
            printText("STREAM_ALARM volume: "+volume+"/"+mBarVolume. getMax());
        }
    }
    private void clearText() {
        mTxtContents.setText("");
    }
    private void printText(String text) {
        mTxtContents.append(text);
        mTxtContents.append("\n");
    }
};
```


AudioManager 应用案例：手持设备音频控制

该案例提供了三个按钮，一个用于音乐的播放，一个用于音乐音量的增加，一个用于音乐音量的减少。该程序运行效果如图 6-4 所示，其详细代码请见配套光盘中的 CH06_4。

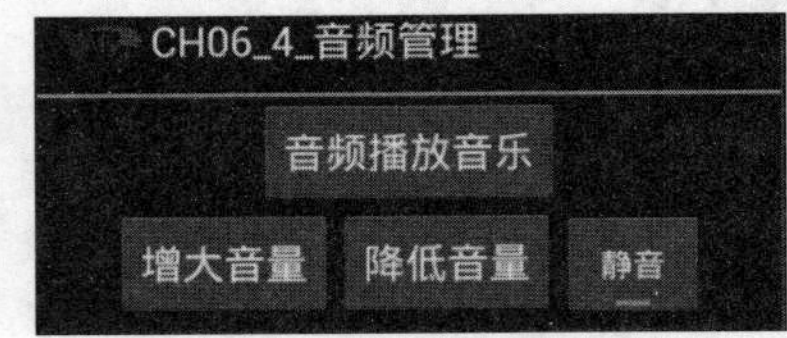

图 6-4　音频管理应用案例

程序的界面布局代码如下：CH06\CH06_4\res\layout\ activity_main.xml。

```
<?xml version="1.0" encoding="utf-8"?>
<LinearLayout xmlns:android="http://schemas.android.com/apk/res/android"
  android:orientation="vertical"
  android:layout_width="fill_parent"
  android:layout_height="fill_parent"
  android:gravity="center_horizontal">
<Button
  android:id="@+id/play"
  android:layout_width="wrap_content"
  android:layout_height="wrap_content"
  android:text="@string/play"  />
<LinearLayout
  android:orientation="horizontal"
  android:layout_width="fill_parent"
  android:layout_height="fill_parent"
  android:gravity="center_horizontal">
<Button
  android:id="@+id/up"
  android:layout_width="wrap_content"
  android:layout_height="wrap_content"
  android:text="@string/up"/>
<Button
  android:id="@+id/down"
  android:layout_width="wrap_content"
  android:layout_height="wrap_content"
```

```
    android:text="@string/down"/>
  <ToggleButton
    android:id="@+id/mute"
    android:layout_width="wrap_content"
    android:layout_height="wrap_content"
    android:textOn="@string/normal"
    android:textOff="@string/mute"/>
  </LinearLayout>
  </LinearLayout>
```

程序的 Java 代码如下： CH06\ \CH06_4\src\com\zigcloud\manager\ AudioTest.java。

```
public class AudioTest extends Activity
{
  Button play, up, down;
  ToggleButton mute;
  AudioManager aManager;
  public void onCreate(Bundle savedInstanceState)
  {
     super.onCreate(savedInstanceState);
     setContentView(R.layout.activity_main);
     //获取系统的音频服务
     aManager=(AudioManager) getSystemService(Service.AUDIO_SERVICE);
     //获取界面中三个按钮和一个ToggleButton控件
     play=(Button) findViewById(R.id.play);
     up=(Button) findViewById(R.id.up);
     down=(Button) findViewById(R.id.down);
     mute=(ToggleButton) findViewById(R.id.mute);
     //为play按钮的单击事件绑定监听器
     play.setOnClickListener(new OnClickListener()
     {
        public void onClick(View source)
        {  //初始化MediaPlayer对象，准备播放音乐
           MediaPlayer mPlayer=MediaPlayer.create(AudioTest.this,R.raw.earth);
           mPlayer.setLooping(true);     //设置循环播放
           mPlayer.start();              //开始播放
        }
     });
     up.setOnClickListener(new OnClickListener()
     {
        public void onClick(View source)
        {   //指定调节音乐的音频，增大音量，而且显示音量图形示意
           aManager.adjustStreamVolume(AudioManager.STREAM_MUSIC,
AudioManager.ADJUST_RAISE, AudioManager.FLAG_SHOW_UI);
        }
     });
     down.setOnClickListener(new OnClickListener()
     {
        public void onClick(View source)
        {  //指定调节音乐的音频，降低音量，而且显示音量图形示意
           aManager.adjustStreamVolume(AudioManager.STREAM_MUSIC,
AudioManager.ADJUST_LOWER, AudioManager.FLAG_SHOW_UI);
```

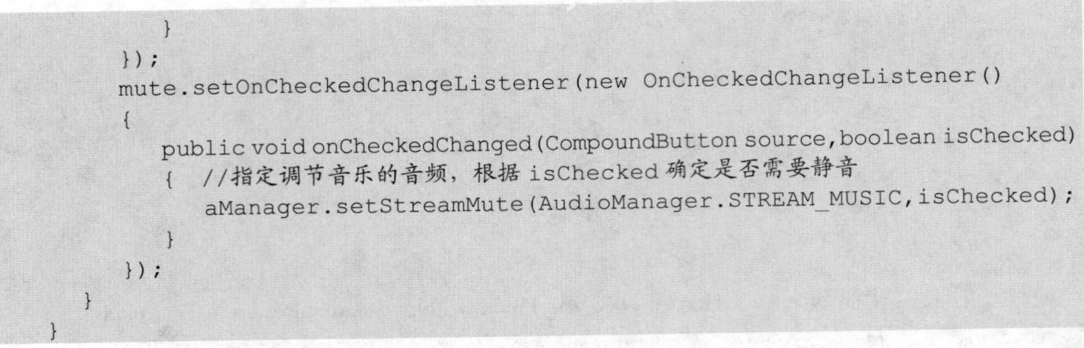

```
            }
        });
        mute.setOnCheckedChangeListener(new OnCheckedChangeListener()
        {
            public void onCheckedChanged(CompoundButton source,boolean isChecked)
            {  //指定调节音乐的音频，根据 isChecked 确定是否需要静音
                aManager.setStreamMute(AudioManager.STREAM_MUSIC,isChecked);
            }
        });
    }
}
```

6.2.4 剪贴板管理器（ClipboardManager）

ClipboardManager 提供到剪贴板的接口，用于设置或获取全局剪贴板中的文本。通过获取系统服务接口，就可以得到 ClipboardManager 对象，如下所示：

```
ClipboardManager service=(ClipboardManager)getSystemService(Context.
CLIPBOARD_SERVICE)
```


在获取了 ClipboardManager 对象后，就可以通过调用其对象的 setText()和 getText()方法设置和获取剪贴板中的文字内容。通过一个简单的例子来学习 ClipboardManager 的使用。源代码清单：CH06\CH06_5\src\sziit\ex\ch06\CH06_5.java。

```
public void onCreate(Bundle savedInstanceState) {
    super.onCreate(savedInstanceState);
    setContentView(R.layout.main);
    ClipboardManager cm=(ClipboardManager)getSystemService(Context.CLIPBOARD_
SERVICE);
    cm.setText("测试剪贴板管理器功能");
}
```

在程序中通过 ClipboardManager 对象的 setText()方法将一段文字放到剪贴板中，然后在通过“粘贴”将剪贴板中的内容放到自定义的 EditText 中。程序运行结果如图 6-5 所示。

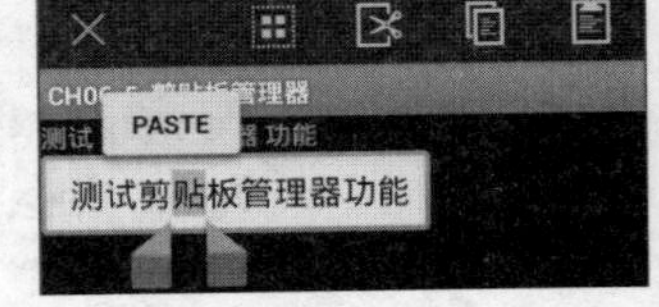

图 6-5　程序运行图

6.2.5 通知管理器（NotificationManager）

NotificationManager 用于通知用户有后台事件发生，其定义于 android.app 包中。当有后台事件发生时，通知管理器会对用户有个提醒。提醒的方式有如下三种：

- 在状态栏中会出现持久的图标，用户可以单击该图标查看通知详情。
- 屏幕开启或者闪烁。
- 通过背景灯闪烁、播放声音或者震动的方式。

首先需要获取 NotificationManager 对象，通过 getSystemService()方法获取通知管理器，然后调用 NotificationManager 对象的 notify()方法可以发送后台事件通知。在 notify()方法中包含两个参数，第一个参数是通知 ID；第二个参数是通知实体。在这里需要注意，通知的发送也是一种预期行为，用户可以在通知发送的时限内取消发送，所以通知的发送也需要使用未决意向对象。

在发送通知时需要定义一个通知实体，也就是 Notification 对象。注意，实体定义使用了三个参数：第一个参数是图标资源 ID，第二个参数是提示文字，第三个参数是通知发送的时间点。下面结合实例进行学习。源代码清单如下：CH06\CH06_6\src\sziit\ ex\ch06\CH06_6.java。

```
    public class CH06_6 extends Activity implements OnClickListener {
       private Button mBtnRegister=null;
       private Button mBtnDo=null;
       private Button mBtnUnregister=null;
       private NotificationManager mService=null;
       private Notification mNotification=null;
       public static final int NOTIFICATION_ID=1;
       public void onCreate(Bundle savedInstanceState) {
       /**第一次创建活动（activity ）时调用*/
           super.onCreate(savedInstanceState);
           setContentView(R.layout.main);
           mBtnRegister=(Button)findViewById(R.id.BTN_REGISTER);
           mBtnDo=(Button)findViewById(R.id.BTN_NOTIFY);
           mBtnUnregister=(Button)findViewById(R.id.BTN_UNREGISTER);
           mBtnRegister.setOnClickListener(this);
           mBtnDo.setOnClickListener(this);
           mBtnUnregister.setOnClickListener(this);
           init();
       }
       private void init() {                 //初始化通知服务
           mService=(NotificationManager)
           getSystemService(Context.NOTIFICATION_SERVICE);
           mNotification=new Notification(R.drawable.tip," 温馨提示",System.
currentTimeMillis() );
           setStates(false);
       }
       public void onClick(View v) {
          switch(v.getId() ) {
             case R.id.BTN_REGISTER: {
                doRegister();
                break;
             }
             case R.id.BTN_NOTIFY: {
                doNotify();
                break;
             }
             case R.id.BTN_UNREGISTER: {
                doUnregister();
                break;
             }
          }
       }
      private void doRegister() {              //注册通知侦听
         Intent notifyIntent=new Intent(this, RemindAct.class);
         notifyIntent.setFlags(Intent.FLAG_ACTIVITY_NEW_TASK);
         PendingIntent contentIntent=PendingIntent.getActivity(this, 0, notify
Intent, 0);
         mNotification.setLatestEventInfo(this.getApplicationContext(),
"新提醒", "已到睡眠时间!", contentIntent);
```

```
        setStates(true);
    }
    private void setStates(boolean isRegistered) {            //设置按钮状态
        mBtnRegister.setEnabled(!isRegistered);
        mBtnUnregister.setEnabled(isRegistered);
        mBtnDo.setEnabled(isRegistered);
    }
    private void doNotify() {               //发出通知
        mService.notify(NOTIFICATION_ID, mNotification);
    }
    private void doUnregister() {           //注销通知侦听
        mService.cancel(NOTIFICATION_ID);
        setStates(false);
    }
};
```

在 doRegister()方法中还指定了通知实体所“呈报”的组件，这里指定 RemindAct 组件展现通知内容。该组件是 Activity 组件，与调用它的组件不存在关联，所以在 Intent 中添加了一个标志：FLAG_ACTIVITY_NEW_TASK。setLatestEventInfo()方法用于设置通知实体的概要信息，同时将通知实体与意向对象绑定。通知实体的概要信息会在单击状态栏图标后展开的界面中显示。最后可以调用 NotificationManager 对象的 cancel()方法取消通知。程序最终运行效果如图 6-6 所示。

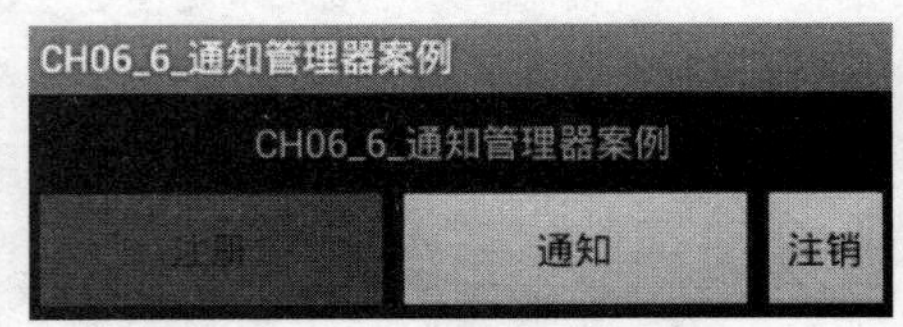

图 6-6　通知管理器案例

6.3　案例——网络诊断案例

与 Windows 平台类似，Android 提供了一些常用的系统服务（如访问网络连接状况、GPS 状态等），这些服务在自己开发应用程序时也会用到，下面的实例实现了一个简单网络诊断工具。具体代码请见配套光盘中的 CH06_7。该程序界面布局如图 6-7 所示。

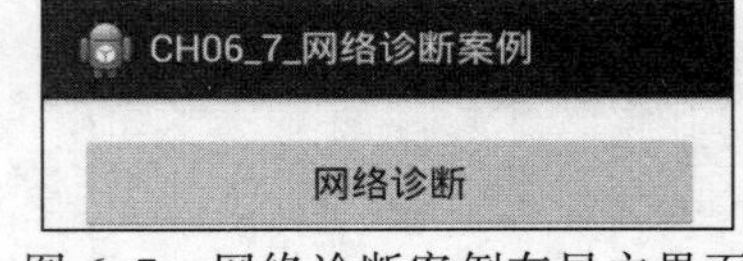

图 6-7　网络诊断案例布局主界面

程序的界面布局代码如下：CH06\ CH06_7\res\layout\activity_main.xml。

```
<RelativeLayout xmlns:android="http://schemas.android.com/apk/res/android"
    xmlns:tools="http://schemas.android.com/tools"
    android:layout_width="match_parent"
    android:layout_height="match_parent"
    android:paddingBottom="@dimen/activity_vertical_margin"
    android:paddingLeft="@dimen/activity_horizontal_margin"
    android:paddingRight="@dimen/activity_horizontal_margin"
    android:paddingTop="@dimen/activity_vertical_margin"
    tools:context=".MainActivity" >
    <Button
        android:id="@+id/btn_diagnostics"
        android:layout_width="wrap_content"
        android:layout_height="wrap_content"
        android:layout_alignParentLeft="true"
```

```
        android:layout_alignParentRight="true"
        android:layout_alignParentTop="true"
        android:text="@string/networkdiagnostics" />
    <TextView
        android:id="@+id/tv_content"
        android:layout_width="wrap_content"
        android:layout_height="wrap_content"
        android:layout_alignLeft="@+id/btn_diagnostics"
        android:layout_alignParentBottom="true"
        android:layout_alignRight="@+id/btn_diagnostics"
        android:layout_below="@+id/btn_diagnostics"/>
</RelativeLayout>
```

程序的 Java 代码如下：CH06\CH06_7\src\com\example\demonetworkdiagnostics\ MainActivity.java。

```
public class MainActivity extends Activity { /*** 诊断网络* */
   private DiagositcsAsyncTask mDiagositcsAsyncTask=new DiagositcsAsyncTask();
   protected void onCreate(Bundle savedInstanceState) {
      super.onCreate(savedInstanceState);
      setContentView(R.layout.activity_main);
      findViewById(R.id.btn_diagnostics).setOnClickListener(new OnClickListener() {

         public void onClick(View v) {
            mDiagositcsAsyncTask=new DiagositcsAsyncTask();
            mDiagositcsAsyncTask.execute();
         }
      });
   }
   public boolean onCreateOptionsMenu(Menu menu) {
      getMenuInflater().inflate(R.menu.main, menu);
                                          //膨胀的菜单，将菜单项添加到行动条
      return true;
   }
   private void setText(final String str){ /*** 设置内容* */
      new Handler().post(new Runnable() {
         public void run() {
            TextView tv_content=(TextView)findViewById(R.id.tv_content);
            tv_content.setText(String.format("%s\r\n%s",tv_content.getText().
toString(), str ));
         }
      });
   }
   /*** 诊断网络* */
   private class DiagositcsAsyncTask extends AsyncTask<String,Integer, String>{
      protected void onPreExecute() {
         super.onPreExecute();
         setText(getResources().getString(R.string.diagnostics_begin));
      }
      protected void onProgressUpdate(Integer... values) {
         super.onProgressUpdate(values);
      }
      protected String doInBackground(String... params) {
```

```
            String res=null;
            try {  //判定是否有网络连接
               boolean
               netState=NetWorkHelper.isNetworkAvailable(MainActivity.this);
               //判断 MOBILE 网络是否可用
               boolean mobileDataState=NetWorkHelper.isMobileDataEnable(Main
Activity. this);
               //检测是否漫游
               boolean netRoamingState=NetWorkHelper.isNetworkRoaming(Main
Activity. this);
               //检测 Wifi 是否可用
               boolean wifiState=NetWorkHelper.isWifiDataEnable(MainActi
vity. this);
               StringBuilder strBuilder=new  StringBuilder();
               strBuilder.append(getResources().getString(R.string.diagnostics_result))
               .append(netState?getResources().getString(R.string.network_
enable):getResources().getString(R.string.network_disable)).append(";")
               .append(mobileDataState?getResources().getString(R.string.
mobiledata_enable):getResources().getString(R.string.mobiledata_disable)).
append(";")
               .append(netRoamingState?getResources().getString(R.string.wifi_
enable):getResources().getString(R.string.wifi_disable)).append(";")
               .append(wifiState?getResources().getString(R.string.roaming_
enable):getResources().getString(R.string.roaming_disable)).append(";");
               res=strBuilder.toString();
            } catch (Exception e) {
               e.printStackTrace();
            }
            return res;
         }
         protected void onPostExecute(String result) {
            super.onPostExecute(result);
            setText(result);
            setText(getResources().getString(R.string.diagnostics_end));
         }
      }
   }
```

程序清单：CH06\CH06_7\src\com\example\demonetworkdiagnostics\https\ APN.java。

```
   package com.example.demonetworkdiagnostics.https;
   public class APN {/*** APN 定义* */
      String apnId;
      String name;
      String numeric;
      String mcc;
      String mnc;
      String apn;
      String user;
      String server;
      String password;
      String proxy;
```

```
    String port;
    String mmsproxy;
    String mmsport;
    String mmsc;
    String authtype;
    String type;
    String current;
    public String getApnId() {
       return apnId;
    }

    public void setApnId(String apnId) {
       this.apnId=apnId;
    }
    public String getName() {
       return name;
    }
    public void setName(String name) {
       this.name=name;
    }
    public String getNumeric() {
       return numeric;
    }
    public void setNumeric(String numeric) {
       this.numeric=numeric;
    }
    public String getMcc() {
       return mcc;
    }
    public void setMcc(String mcc) {
       this.mcc=mcc;
    }
    public String getMnc() {
       return mnc;
    }
    public void setMnc(String mnc) {
       this.mnc=mnc;
    }
    public String getApn() {
       return apn;
    }
    public void setApn(String apn) {
       this.apn=apn;
    }
    public String getUser() {
       return user;
    }
    public void setUser(String user) {
       this.user=user;
    }
```

```
public String getServer() {
   return server;
}
public void setServer(String server) {
   this.server=server;
}
public String getPassword() {
   return password;
}
public void setPassword(String password) {
   this.password=password;
}
public String getProxy() {
   return proxy;
}
public void setProxy(String proxy) {
   this.proxy=proxy;
}
public String getPort() {
   return port;
}
public void setPort(String port) {
   this.port=port;
}
public String getMmsproxy() {
   return mmsproxy;
}
public void setMmsproxy(String mmsproxy) {
   this.mmsproxy=mmsproxy;
}
public String getMmsport() {
   return mmsport;
}
public void setMmsport(String mmsport) {
   this.mmsport=mmsport;
}
public String getMmsc() {
   return mmsc;
}
public void setMmsc(String mmsc) {
   this.mmsc=mmsc;
}
public String getAuthtype() {
   return authtype;
}
public void setAuthtype(String authtype) {
   this.authtype=authtype;
}
public String getType() {
   return type;
```

```
    }
    public void setType(String type) {
        this.type=type;
    }
    public String getCurrent() {
        return current;
    }
    public void setCurrent(String current) {
        this.current=current;
    }
}
```

程序清单：CH06\CH06_7\src\com\example\demonetworkdiagnostics\https\ APNManager.java。

```
public class APNManager {/*** APN 控制**/
    private ContentResolver resolver;
    private static final Uri PREFERRED_APN_URI=Uri.parse("content://telephony/
carriers/preferapn");
    private static final Uri APN_TABLE_URI=Uri.parse("content:// telephony/
carriers");
    private TelephonyManager tm;
    private Context mContext;
    private static APNManager apnManager=null;
    public static APNManager getInstance(Context context) {
        if(apnManager!=null) {
            apnManager=new APNManager(context);
        }
        return apnManager;
    }
    private APNManager(Context context) {
        resolver=context.getContentResolver();
        mContext=context;
        tm=(TelephonyManager) context.getSystemService(Context.TELEPHONY_SERVICE);
    }
    /*** 判断一个 apn 是否存在，存在返回 id*
     * @param apnNode
     * @return
     */
    @SuppressWarnings("unused")
    public int isApnExisted(APN apnNode) {
        int apnId = -1;
        Cursor mCursor=resolver.query(APN_TABLE_URI,null,null,null,null);
        while(mCursor != null && mCursor.moveToNext()) {
            apnId=mCursor.getShort(mCursor.getColumnIndex("_id"));
            String name=mCursor.getString(mCursor.getColumnIndex("name"));
            String apn=mCursor.getString(mCursor.getColumnIndex("apn"));
            String type=mCursor.getString(mCursor.getColumnIndex("type"));
            String proxy=mCursor.getString(mCursor.getColumnIndex("proxy"));
            String port=mCursor.getString(mCursor.getColumnIndex("port"));
            String current=mCursor.getString(mCursor.getColumnIndex("current"));
            String mcc=mCursor.getString(mCursor.getColumnIndex("mcc"));
            String mnc=mCursor.getString(mCursor.getColumnIndex("mnc"));
```

```
            String numeric=mCursor.getString(mCursor.getColumnIndex("numeric"));
            Log.e("isApnExisted", "info:" + apnId + "_" + name + "_" + apn +
"_" + type + "_" + current + "_" + proxy);// 遍历了所有的apn
            if (/* apnNode.getName().equals(name) */(apnNode.getApn().equals(apn)
&& apnNode.getMcc().equals(mcc)&& apnNode.getMnc().equals(mnc) && apnNode.
getNumeric()   .equals(numeric))&& (type == null || "default".equals(type) ||
"".equals(type)))
                        // || (apnNode.getApn().equals(apn)&& "".equals(proxy)
&&"".equals(port))
            {
               return apnId;
            } else {
               apnId = -1;
            }
         }
         return apnId;
      }

      /*** 设置默认的apn
       * @param apnId
       * @return
       */
      public boolean setDefaultApn(int apnId) {
         boolean res=false;
         ContentValues values=new ContentValues();
         values.put("apn_id", apnId);
         try {
            resolver.update(PREFERRED_APN_URI, values, null, null);
            Cursor c=resolver.query(PREFERRED_APN_URI, new String[] { "name",
"apn" }, "_id="+apnId, null, null);
            if(c!=null) {
               res=true;
               c.close();
            }
         } catch (SQLException e) {
            e.printStackTrace();
         }
         return res;
      }
      public void deleteApn() {/*** 删除所有apn*/
         resolver.delete(APN_TABLE_URI, null, null);
      }
      @SuppressWarnings("unused")
      public APN getDefaultAPN() {
         String id="";
         String apn="";
         String proxy="";
         String name="";
         String port="";
         String type="";
```

```
        String mcc="";
        String mnc="";
        String numeric="";
        APN apnNode=new APN();
        Cursor mCursor=resolver.query(PREFERRED_APN_URI,null,null, null,null);
        if(mCursor==null) {
            throw new Exception("不存在喜欢的 APN ");
            return null;
        }
        while(mCursor != null && mCursor.moveToNext()) {
            id=mCursor.getString(mCursor.getColumnIndex("_id"));
            name=mCursor.getString(mCursor.getColumnIndex("name"));
            apn=mCursor.getString(mCursor.getColumnIndex("apn")).toLowerCase();
            proxy=mCursor.getString(mCursor.getColumnIndex("proxy"));
            port=mCursor.getString(mCursor.getColumnIndex("port"));
            mcc=mCursor.getString(mCursor.getColumnIndex("mcc"));
            mnc=mCursor.getString(mCursor.getColumnIndex("mnc"));
            numeric=mCursor.getString(mCursor.getColumnIndex("numeric"));
            Log.d("getDefaultAPN", "default Apn info:" + id + "_" + name + "_"
+ apn + "_" + proxy + "_" + proxy);
        }
        apnNode.setName(name);
        apnNode.setApn(apn);
        apnNode.setProxy(proxy);
        apnNode.setPort(port);
        apnNode.setMcc(mcc);
        apnNode.setMnc(mnc);
        apnNode.setNumeric(numeric);
        return apnNode;
    }
    public int getDefaultNetworkType() {
        int networkType = -1;
        ConnectivityManager connectivity=(ConnectivityManager) mContext
.getSystemService(Context.CONNECTIVITY_SERVICE);
        if(connectivity==null) {
        } else {
            // WIFI 网络优先
            NetworkInfo wifiNetworkInfo=connectivity
                    .getNetworkInfo(ConnectivityManager.TYPE_WIFI);
           if(wifiNetworkInfo!=null&& wifiNetworkInfo.getState()==NetworkInfo.
State.CONNECTED) {
                return ConnectivityManager.TYPE_WIFI;
            }
            NetworkInfo[] info=connectivity.getAllNetworkInfo();
            for(int i=0; i<info.length; i++) {
                if(info[i].getState()==NetworkInfo.State.CONNECTED) {
                    networkType=info[i].getType();//使用第一个可用的网络
                    break;
                }
            }
```

```
    }
    return networkType;
}
public int InsetAPN() {                    //添加一个 APN
    APN checkApn=new APN();
    checkApn.setName("爱家物联专用接口");
    checkApn.setApn("ctnet");
    //checkApn.setProxy("10.0.0.200");
    //checkApn.setPort("80");
    checkApn.setUser("qdaj@qdaj.vpdn.sd");
    checkApn.setPassword("123456");
    checkApn.setMcc(getMCC());
    checkApn.setMnc(getMNC());
    checkApn.setNumeric(getSimOperator());
    return addNewApn(checkApn);
}
/*** 增加新的 apn
 * @param apnNode
 * @return
 */
private int addNewApn(APN apnNode) {
    int apnId = -1;
    ContentValues values=new ContentValues();
    values.put("name", apnNode.getName());
    values.put("apn", apnNode.getApn());
    values.put("proxy", apnNode.getProxy());
    values.put("port", apnNode.getPort());
    values.put("user", apnNode.getUser());
    values.put("password", apnNode.getPassword());
    values.put("mcc", apnNode.getMcc());
    values.put("mnc", apnNode.getMnc());
    values.put("numeric", apnNode.getNumeric());
    Cursor c=null;
    try {
        Uri newRow=resolver.insert(APN_TABLE_URI, values);
        if(newRow!=null) {
            c=resolver.query(newRow, null, null, null, null);
            int idindex=c.getColumnIndex("_id");
            c.moveToFirst();
            apnId=c.getShort(idindex);
            Log.d("Robert","New ID: "+apnId+":Inserting new APN succeeded!");
        }
    } catch (SQLException e) {
        e.printStackTrace();
    }
    if(c!=null)
        c.close();
    return apnId;
}
private String getMCC() {
```

```
        String numeric=tm.getSimOperator();
        String mcc=numeric.substring(0, 3);
        Log.i("MCC  is", mcc);
        return mcc;
    }
    private String getMNC() {
        String numeric=tm.getSimOperator();
        String mnc=numeric.substring(3, numeric.length());
        Log.i("MNC  is", mnc);
        return mnc;
    }
    private String getSimOperator() {
        String SimOperator=tm.getSimOperator();
        return SimOperator;
    }
    public String matchAPN(String currentName) {
        if ("".equals(currentName) || null==currentName) {
            return "";
        }
        currentName=currentName.toLowerCase();
        if (currentName.startsWith("cmnet") || currentName.startsWith("CMNET"))
            return "cmnet";
        else if (currentName.startsWith("cmwap")|| currentName.startsWith("CMWAP"))
            return "cmwap";
        else if (currentName.startsWith("3gwap")|| currentName.startsWith("3GWAP"))
            return "3gwap";
        else if (currentName.startsWith("3gnet")|| currentName.startsWith("3GNET"))
            return "3gnet";
        else if (currentName.startsWith("uninet")|| currentName.startsWith("UNINET"))
            return "uninet";
        else if (currentName.startsWith("uniwap")|| currentName.startsWith("UNIWAP"))
            return "uniwap";
        else if (currentName.startsWith("default")|| currentName.startsWith("DEFAULT"))
            return "default";
        else
            return "";
    }
    /*** 获取Apn列表
     * @param context
     * @return
     */
    public List<APN> getAPNList() {
        String tag="Main.getAPNList()";
        String projection[]={ "_id,apn,type,current" };
        Cursor cr=mContext.getContentResolver().query(APN_TABLE_URI, projection,
null,null,null);
        List<APN> list=new ArrayList<APN>();
        while (cr!=null && cr.moveToNext()) {
            Log.d(tag, cr.getString(cr.getColumnIndex("_id")) + "  "
                    + cr.getString(cr.getColumnIndex("apn")) + "  "
```

```
                    + cr.getString(cr.getColumnIndex("type")) + "  "
                    + cr.getString(cr.getColumnIndex("current")));
            APN a=new APN();
            a.apnId=cr.getString(cr.getColumnIndex("_id"));
            a.apn=cr.getString(cr.getColumnIndex("apn"));
            a.type=cr.getString(cr.getColumnIndex("type"));
            list.add(a);
        }
        if(cr!=null)
            cr.close();
        return list;
    }
}
```

程序清单：CH06\CH06_7\src\com\example\demonetworkdiagnostics\https\ NetWorkHelper.java。

```
public class NetWorkHelper {
    private static String LOG_TAG="NetWorkHelper";
    public static Uri uri=Uri.parse("content://telephony/carriers");
    public static boolean isNetworkAvailable(Context context) {
    /*** 判断是否有网络连接*/
        ConnectivityManager connectivity=(ConnectivityManager) context
.getSystemService(Context.CONNECTIVITY_SERVICE);
        if(connectivity==null) {
            Log.w(LOG_TAG, "无法获得连接管理器");
        } else {
            NetworkInfo[] info=connectivity.getAllNetworkInfo();
            if(info!=null) {
                for(int i=0; i<info.length; i++) {
                    if (info[i].isAvailable()) {
                        Log.d(LOG_TAG, "网络可用");
                        return true;
                    }
                }
            }
        }
        Log.d(LOG_TAG, "network is not available");
        return false;
    }
    public static boolean checkNetState(Context context){
        boolean netstate=false;
        ConnectivityManager     connectivity=(ConnectivityManager)context.
getSystemService(Context.CONNECTIVITY_SERVICE);
        if(connectivity!=null)
        {
            NetworkInfo[] info=connectivity.getAllNetworkInfo();
            if(info!=null) {
                for(int i=0; i<info.length; i++)
                {
                if(info[i].getState()==NetworkInfo.State.CONNECTED)
                    {
                        netstate=true;
```

```
                    break;
                }
            }
        }
    }
    return netstate;
  }
  public static boolean isNetworkRoaming(Context context) {
  /*** 判断网络是否为漫游*/
    ConnectivityManager connectivity=(ConnectivityManager) context
.getSystemService(Context.CONNECTIVITY_SERVICE);
    if(connectivity==null) {
      Log.w(LOG_TAG, "无法获得连接管理器!");
    } else {
      NetworkInfo info=connectivity.getActiveNetworkInfo();
      if(info != null&& info.getType() == ConnectivityManager.TYPE_MOBILE)
{TelephonyManager tm=(TelephonyManager) context.getSystemService(Context.
TELEPHONY_SERVICE);
        if (tm != null && tm.isNetworkRoaming()) {
          Log.d(LOG_TAG, "网络正在漫游!");
          return true;
        } else {
          Log.d(LOG_TAG, "网络停止漫游!");
        }
      } else {
        Log.d(LOG_TAG, "移动网络不可用! ");
      }
    }
    return false;
  }
  /** * 判断 MOBILE 网络是否可用
   * @param context
   * @return
   * @throws Exception
   */
  public static boolean isMobileDataEnable(Context context) throws Exception {
    ConnectivityManager connectivityManager=(ConnectivityManager) context
.getSystemService(Context.CONNECTIVITY_SERVICE);
    boolean isMobileDataEnable=false;
    isMobileDataEnable=connectivityManager.getNetworkInfo(
ConnectivityManager.TYPE_MOBILE).isConnectedOrConnecting();
    return isMobileDataEnable;
  }
  /**
   * 判断 wifi 是否可用
   * @param context
   * @return
   * @throws Exception
   */
  public static boolean isWifiDataEnable(Context context) throws Exception {
```

```
        ConnectivityManager connectivityManager=(ConnectivityManager) context
.getSystemService(Context.CONNECTIVITY_SERVICE);
        boolean isWifiDataEnable=false;
        isWifiDataEnable=connectivityManager.getNetworkInfo(
ConnectivityManager.TYPE_WIFI).isConnectedOrConnecting();
        return isWifiDataEnable;
    }
    /*** 设置Mobile网络开关
     * @param context
     * @param enabled
     * @throws Exception
     */
    public static void setMobileDataEnabled(Context context, boolean enabled)
throws Exception {
        APNManager apnManager=APNManager.getInstance(context);
        List<APN> list=apnManager.getAPNList();
        if (enabled) {
            for(APN apn : list) {
                ContentValues cv=new ContentValues();
                cv.put("apn", apnManager.matchAPN(apn.apn));
                cv.put("type", apnManager.matchAPN(apn.type));
                context.getContentResolver().update(uri, cv, "_id=?",new String[]
{ apn.apnId });
            }
        } else {
            for (APN apn : list) {
                ContentValues cv=new ContentValues();
                cv.put("apn", apnManager.matchAPN(apn.apn)+"mdev");
                cv.put("type", apnManager.matchAPN(apn.type)+"mdev");
                context.getContentResolver().update(uri, cv, "_id=?",new String[]
{ apn.apnId });
            }
        }
    }
}
```

该程序的运行效果如图 6-8 所示。

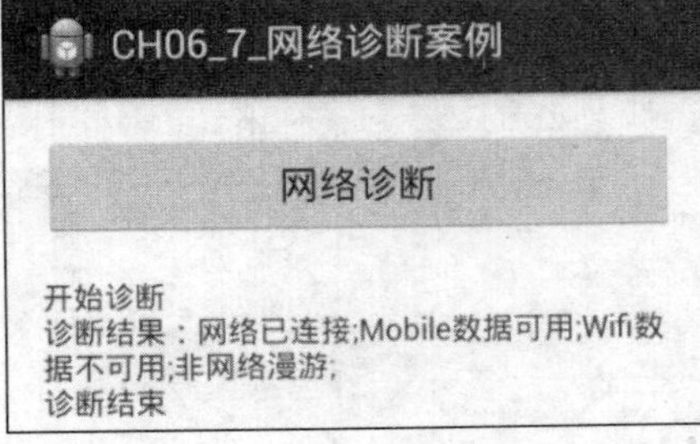

图 6-8　网络诊断案例运行主界面

6.4 知识扩展

6.4.1 电话管理器（TelephonyManager）

TelephonyManager 是一个管理手机通话状况、电话网络信息的服务类，该类提供了大量的 getXxx()方法获取电话网络的相关信息。

在程序中获取 TelephonyManager 非常简单，只需调用如下代码即可：

```
    TelephonyManager  tManager=(TelephonyManager)getSystemService(Context.
TELEPHONY_SERVICE);
```

6.4.2 短信管理器（SmsManager）

SmsManager 提供了系列 sendXxxMessage()方法用于发送短信。实现这个功能非常简单，直接调用 sendTextMessage()方法即可。

6.5 本章小结

本章讲述 Android 主要的几个系统服务，活动管理器（ActivityManager）、警报管理器（AlarmManager）、音频管理器（AudioManager）、剪贴板管理器（ClipboardManager）、通知管理器（NotificationManager）的概念、作用及实现。

6.6 强化练习

1．填空题

（1）（　　）是对所有运行中的 Activity 组件进行管理，在 android.app 包中。

（2）通过配置信息接口的 reqNavigation 属性可以获得当前设备的导航方式，导航方式类型包括（　　）、（　　）、（　　）和（　　）。

（3）（　　）允许用户预定自定义应用程序的运行时间。

（4）设置警报主要分为两种：（　　）和（　　）。

（5）（　　）提供到剪贴板的接口，用于设置或获取全局剪贴板中的文本。

（6）（　　）又称为通知管理器，用于通知用户有后台事件发生，其定义在 android.app 包中。

2．问答题、操作题或编程题

（1）编程实现通过调用 AudioManager 对象的方法实现对音量的设置和管理。

（2）编程实现设置和获取剪贴板中的文字内容。

（3）编程实现通知管理。

综合项目实训篇

"纸上得来终觉浅，绝知此事要躬行。"本部分在Android核心理论知识的基础上，通过3个企业真实的、涉及当前热门的物联网技术和移动互联技术等方面的典型综合项目，基于软件工程化思想，按照企业项目实施的关键结点，详细论述每个项目的设计与实现，以提高学生综合应用多种核心知识和技能，解决复杂工程问题的素养和能力。

本篇包含以下3章：

第7章　基于Android智能家居系统项目

第8章　Android校园通项目

第9章　基于Android智能仓储系统项目

第7章

基于 Android 智能家居系统项目

学习目标：

- 掌握软件界面的设计和实现；
- 掌握页面与页面之间数据交互的实现；
- 掌握数据库的设计和实现。

7.1 项目概述

物联网智能家居系统是基于 RFID(Radio Frequency Identification，射频识别)技术、Zigbee 技术、Wi-Fi 技术等的可用于真实项目的系统。整个系统具有八大子系统：环境监控系统、智能照明系统、智能遮阳系统、安防系统、智能门禁系统、视频监控系统、电器控制系统和远程控制系统。

无论何时何地，都可以通过手持设备或者计算机控制家电和视频监控，实现灯光开关控制、窗帘控制、视频监控等功能，同时还支持防盗报警、燃气报警、烟雾报警等功能。集成了目前智能家居中众多声光电传感器，能够很好地展示现代家居生活的舒适化和智能化。

7.1.1 平台简介

实训台实物图如图 7-1 所示。

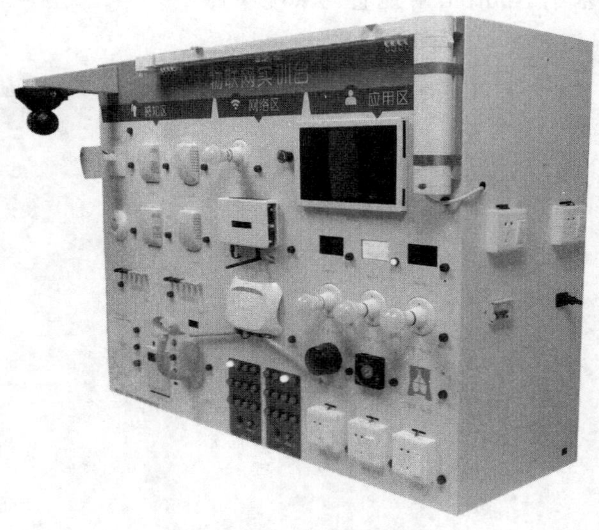

图 7-1 实训台实物图

智能家居实训台具有八大系统：环境监控系统、智能照明系统、智能遮阳系统、安防系统、智能门禁系统、视频监控系统、电器控制系统和远程控制系统。

智能家居实训台高度结合了物联网工程技术与行业体系架构，还原了行业真实环境，将感知层、网络层和应用层三个区域分开、区域清晰；预留扩展接口，方便用户二次开发调试；烧写接口集中管控方便烧写；集成远程教学功能；智能断电系统，过电流可自动切断电源。

7.1.2 硬件资源

系统分为三个区：感知区、网络区和应用区。

1. **感知区**

（1）视频部分：IP Camera。

（2）传感器部分：红外探测器、燃气探测器、温湿度探测器、光照探测器、烟雾探测器和空气质量探测器。

（3）扩展的数据采器部分：数据采集器 I 和数据采集器 II。（可以外接传感器，支持模拟量、开关量和串行数据的输入）。

（4）门装置部分：门磁、门锁、指纹采集器和 RFID 读卡器。

（5）扫描码采集部分：一维扫描枪。

实训台硬件简介如图 7-2 所示。

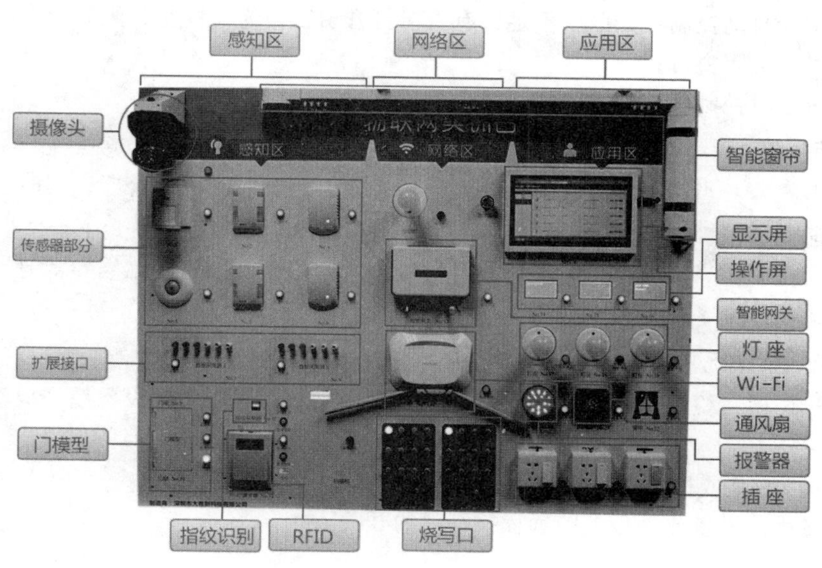

图 7-2 实训台硬件简介图

2. **网络区**

（1）辅助部分：照光灯和急停按钮。

（2）网络通信部分：智能网关和无线 AP（Access Point，访问结点）。

（3）程序下载部分：程序下载板 1 和程序下载板 2。

3. 应用区

（1）显示部分：平板电脑、字符液晶 1、字符液晶 2 和字符液晶 3。

（2）控制部分：灯座 1、灯座 2、灯座 3、报警器、通风扇、窗帘、插座 1、插座 2、插座 3。

7.2 项目设计

7.2.1 项目总体功能需求

1. 环境监控系统

在手持设备或平板电脑上能查看到：温度、温度、光照度、空气质量、燃气、红外、烟雾等传感器的状况。

2. 智能照明系统

（1）使用手持设备或平板电脑可以远程控制灯光的开启和关闭。

（2）灯光的自动控制：光照度不足，有人时打开灯；光照度足，无人或有人时及光照度足，无人时，关闭灯。

（3）灯光定时控制。

3. 智能遮阳系统

（1）使用手持设备或平板电脑可以远程控制窗帘的开启和关闭。

（2）窗帘的自动控制：光照度不足，有人时打开窗帘；光照度足，无人或有人时及光照度足，无人时，关闭窗帘；

（3）窗帘定时控制。

4. 安防系统

（1）防盗报警：当家里没有人，系统检测到有人入侵时，启动报警器，同时，手持设备或平板电脑能显示报警信息。

（2）防灾报警：当家里有燃气和烟雾异常时，启动报警器，当是燃气异常时，打开通风设备，同时，手持设备或平板电脑能显示报警信息。

5. 智能门禁系统

（1）通过门磁，检测门的开关状况，并以指示灯显示，同时手持设备或平板电脑能显示门的状况。

（2）通过 RFID 或指纹设备能控制门的开关。

（3）通过手持设备或平板电脑远程控制门的开关。

6. 视频监控

手持设备或平板电脑远程查看到家里的状况。

7. 电器控制系统

（1）在手持设备或平板电脑远程能查看到：灯、风扇、热水器等用电设备的状况。

（2）在手持设备远程能控制：风扇、热水器等用电设备的开关。

8. **情景模式**

（1）起床模式：清晨的卧室，窗帘拉开，当光线不足时，灯光自动开启。

（2）离家模式：早餐后，该上班了，一个按键即可关闭所有的电器，同时家庭中安防系统自动开启，简单、省时。

（3）回家模式：到家后，灯感应开启，启动回家模式，关闭防盗安防，客厅电动窗帘自动拉开。

（4）晚餐模式：晚餐时间到了，“就餐”灯光效果启动，窗帘自动打开，享受美好的晚餐。

（5）家庭影院模式：用餐完毕，欣赏一场电影吧，只需一个按键，瞬间营造一个华丽的电影院。

（6）睡眠模式：该休息了，轻按“休息”场景键，开启睡眠模式，灯光调节到合适的亮度，窗帘缓缓合上，开始您美梦时刻。

7.2.2 项目总体设计

现在全球在进行信息化建设，而信息化建设最终能否落地，移动终端上的应用开发将起到决定性作用，Android是目前市场占有率最高的移动设备操作系统，同时它的操作系统是开源的，这样有利于其上应用程序的开发。因此，本系统的应用程序选择在Android操作系统上进行开发。

从安全性和便利性方面考虑，本系统将采用云与端的方式进行开发。

本系统采用常规的用户与密码的方式，登录进入系统，在系统主页面中有六个采用扁平化设计的功能选项：情景模式、我的房间、家用电器、视频监控、系统设置和更多。

根据人的一天的常规作息规律，在“情景模式”中包括：起床模式、离家模式、回家模式、就餐模式、影院模式、睡眠模式和“+”，“+”功能主要是考虑功能的扩展，用户可以根据个人的喜好添加新的模式，同时也可以对模式进行删除操作。

根据家庭的布局，在“我的房间”中包括：客厅、主卧、次卧、书房、厨房、卫生间、阳台、车库和“+”；“+”功能主要是考虑功能的扩展，用户可以根据个人家庭的情况添加新的房间，同时也可以对房间进行删除操作。进入到具体的房间后，会显示该房间中所有的电器设备列表包含传感器，并可以查看到简单的设备状况信息；点击具体的电器后，可以查看电器具体信息，如果是可控设备，则可以对设备进行控制。

根据家庭中电器的种类，在“家用电器”中包括：环境监控、灯光控制、门窗控制、电风扇、电视、空调和“+”；“+”功能主要是考虑功能的扩展，用户可以根据个人家庭的情况添加新的设备，同时也可以对设备进行删除操作。选择具体的功能模块后，会显示家庭中各种电器的列表及简单设备信息，点击具体的电器后，可以查看电器具体信息，如果是可控设备，则可以对设备进行控制。

根据家庭中所安装位置的不同，在“视频监控”中包括：门厅监控、客厅监控、书房监控、车库监控和“+”；“+”功能主要是考虑功能的扩展，用户可以根据个人家庭的情况添加新的房间，同时也可以对房间进行删除操作。选择具体的房间后，可以查看到所安装的摄像头列表，选中具体的摄像头时，就可以查看到具体的视频监控信息。

“系统设备”中主要包括：用户管理、密码管理等其他功能。

“更多”这个选项用于后期功能的扩展。项目整体设计框架如图 7-3 所示。

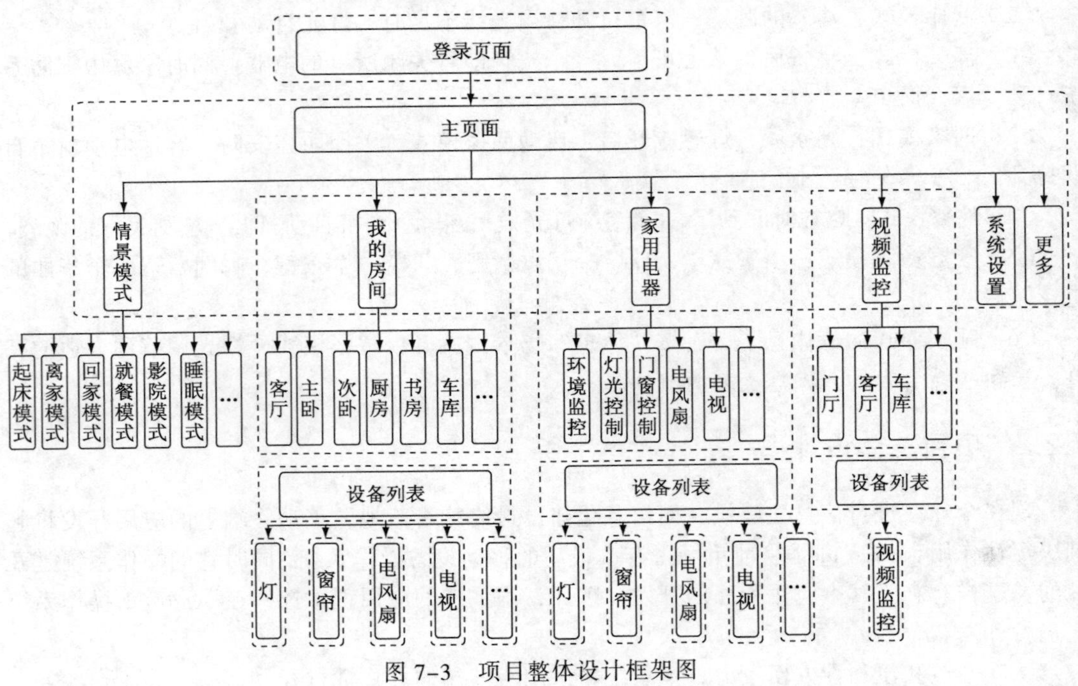

图 7-3　项目整体设计框架图

7.3　必备的技术和知识点

本章必备的技术和知识点包括：

（1）界面编程与视图组件。

（2）布局管理器。

（3）TextView 及其子类。

（4）ImageView 及其子类。

（5）AdapterView 及其子类。

（6）ViewAnimator 及其子类。

（7）对话框和菜单。

（8）基于监听的事件处理。

（9）基于回调的事件处理。

（10）响应系统设置的事件。

（11）Handle 消息传递机制。

（12）基于 TCP 的网络通信。

（13）使用 URL 访问网络资源。

（14）使用 WebServer 进行网络编程。

（15）数据存储与访问。

（16）嵌入式关系型 SQLite 数据库存储数据。

（17）视频采集。

7.4 项目实施

7.4.1 登录页面

实现账号和密码输入文本框、实现登录按键功能、实现密码记录功能和自动登录功能。页面布局如图 7-4 所示。

图 7-4 登录页面图

当用户输入账号和密码，点击登录按键时，系统读取账号和密码文本框中的信息，并把账号和密码信息经网络传输给智能网关，与智能网关中的数据库中的账号和密码数据信息进行比对，如果账号和密码同时存在于智能网关中的数据库中，则允许用户登录系统。如果账号或密码不正确，则根据返回的标记码的不同，提示“账号不存在”或“密码不正确”。

当用户选中“保存密码”时，系统会把账号和密码保存在相关的配置文件中，当下次再登录系统时，则不需要再次输入账号和密码。如果用户只选中“自动登录”时，则系统会保存账号和密码，下次自动登录系统。程序的界面布局代码如下：

程序清单：CH07\CHOME\src\com\zigcloud\smarthome\activity\LoginActivity.java。

```
public class LoginActivity extends BaseActivity{ /*** 用户登录* */
  public void onCreate(Bundle savedInstanceState) {
    super.onCreate(savedInstanceState);
    setContentView(R.layout.activity_login);
    MainApplication.getInstance().addActivity(this);
    findViewById(R.id.btn_login).setOnClickListener(new View.OnClickListener() {
        public void onClick(View v) {
           startActivity(new Intent(LoginActivity.this,MainActivity.class));
        }
     });
  }
  public void onBackPressed() {
     super.onBackPressed();
  }
  public boolean onKeyDown(int keyCode, KeyEvent event) {
     if(keyCode==KeyEvent.KEYCODE_BACK)
     {
         exitBy2Click();
     }
     return false;
  }
  private static Boolean isExit=false;
  private void exitBy2Click() {/*** 双击后退按钮退出* */
     Timer tExit=null;
```

```
        if(isExit==false) {
            isExit=true;
            Toast.makeText(this, getResources().getString(R.string.exit_dialog_
message), Toast.LENGTH_SHORT).show();
            tExit=new Timer();
            tExit.schedule(new TimerTask() {
                public void run() {
                    isExit=false;
                }
            }, 2000);
        } else {
            MainApplication.getInstance().exit();
        }
    }
}
```

7.4.2 主页面

实现“返回”和六个示图功能。页面布局如图 7-5 所示。

“返回”功能的作用是返回上一级页面；“情景模式”示图的作用是进入到情景模式页面。其他五张示图功能类似。程序的界面布局代码如下：

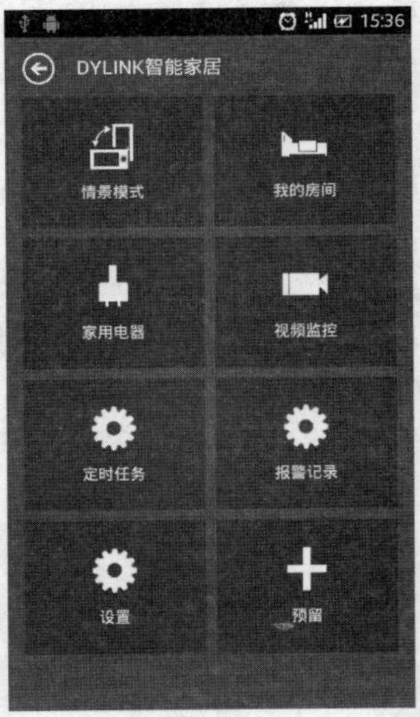

图 7-5　主页面图

程序清单：CH07\CHOME\res\layout\activity_main.xml。

```
<?xml version="1.0" encoding="utf-8"?>
<RelativeLayout xmlns:android="http:// schemas.
android.com/apk/res/android"
    android:layout_width="fill_parent"
    android:layout_height="fill_parent"
    android:background="@drawable/main_default_bg" >
    <include android:id="@id/bt_createtask_title_
layout"
        layout="@layout/activity_common_title_bar" />
    <RelativeLayout
        android:id="@+id/RelativeLayout1"
        android:layout_width="match_parent"
        android:layout_height="match_parent"
        android:layout_below="@id/bt_createtask_title_layout"
        android:layout_centerHorizontal="true"
        android:layout_marginLeft="15dp"
        android:layout_marginRight="15dp"
        android:orientation="vertical"
        android:visibility="visible" >
        <LinearLayout
            android:id="@+id/LinearLayout1"
            android:layout_width="match_parent"
            android:layout_height="match_parent"
```

```
            android:layout_alignParentTop="true"
            android:orientation="vertical" >
            <GridView
              android:id="@+id/gdv_main"
              android:layout_width="match_parent"
              android:layout_height="wrap_content"
              android:listSelector="@drawable/selector_list"
              android:horizontalSpacing="10dp"
              android:numColumns="2"
              android:verticalSpacing="10dp" >
          </GridView>
            <include
               android:id="@+id/view_loading_error"
               android:layout_width="match_parent"
               android:layout_height="match_parent"
               layout="@layout/view_loading_error"
               android:visibility="gone" />
            <include
               android:id="@+id/view_loading"
               android:layout_width="match_parent"
               android:layout_height="match_parent"
               layout="@layout/view_loading"
               android:visibility="gone" />
        </LinearLayout>
    </RelativeLayout>
</RelativeLayout>
```

程序清单：CHOME\src\com\zigcloud\smarthome\activity\MainActivity.java。

```
public class MainActivity extends BaseActivity{ /***主界面 **/
   private TextView titlebar_left;
   private TextView titlebar_title;
   public void onCreate(Bundle savedInstanceState) {
      super.onCreate(savedInstanceState);
      setContentView(R.layout.activity_main);
      MainApplication.getInstance().addActivity(this);
      titlebar_left=(TextView) findViewById(R.id.titlebar_left);
      titlebar_left.setOnClickListener(new View.OnClickListener() {
         public void onClick(View v) {
            onBackPressed();
         }
      });
      titlebar_title=(TextView) findViewById(R.id.titlebar_title);
      titlebar_title.setText(R.string.app_name);
      GridView gridView=(GridView) findViewById(R.id.gdv_main);
      ArrayList<HashMap<String,Object>> lstImageItem=new ArrayList <HashMap
<String, Object>>();
     HashMap<String, Object> map = new HashMap<String, Object>();
      map.put("ItemImage", R.drawable.icon_model);
```

```
        map.put("ItemText", "情景模式");
        lstImageItem.add(map);
        map=new HashMap<String, Object>();
        map.put("ItemImage", R.drawable.icon_room);
        map.put("ItemText", "我的房间");
        lstImageItem.add(map);
        map=new HashMap<String, Object>();
        map.put("ItemImage", R.drawable.icon_electric);
        map.put("ItemText", "家用电器");
        lstImageItem.add(map);
        map=new HashMap<String, Object>();
        map.put("ItemImage", R.drawable.icon_camera);
        map.put("ItemText", "视频监控");
        lstImageItem.add(map);
        map=new HashMap<String, Object>();
        map.put("ItemImage", R.drawable.icon_setting);
        map.put("ItemText", "定时任务");
        lstImageItem.add(map);
        map=new HashMap<String, Object>();
        map.put("ItemImage", R.drawable.icon_setting);
        map.put("ItemText", "报警记录");
        lstImageItem.add(map);
        map=new HashMap<String, Object>();
        map.put("ItemImage", R.drawable.icon_setting);
        map.put("ItemText", "设置");
        lstImageItem.add(map);
        map=new HashMap<String, Object>();
        map.put("ItemImage", R.drawable.icon_add);
        map.put("ItemText", "预留");
        lstImageItem.add(map);
        SimpleAdapter saImageItems=new SimpleAdapter(this, lstImageItem,
R.layout.view_squared_item, new String[] {"ItemImage","ItemText"}, new int[]
{R.id.itemImage,R.id.itemText});
        gridView.setAdapter(saImageItems);
        gridView.setOnItemClickListener(new OnItemClickListener() {
            public void onItemClick(AdapterView<?>arg0,View arg1,int arg2,long arg3){
                switch(arg2){
                    case 0:
                        startActivity(new
                        Intent(MainActivity.this,ModelListActivity.class));
                        break;
                    case 1:
                        startActivity(new
                        Intent(MainActivity.this,RoomListActivity.class));
                        break;
                }
            }
        });
        gridView.setSelector(new ColorDrawable(Color.TRANSPARENT));
    }
    public boolean onKeyDown(int keyCode, KeyEvent event){
```

```
        if(keyCode==KeyEvent.KEYCODE_BACK)
        {
            exitBy2Click();
        }
        return false;
    }
    private static Boolean isExit=false;
    private void exitBy2Click() {
        Timer tExit=null;
        if(isExit==false) {
            isExit=true;
            Toast.makeText(this,getResources().getString(R.string.exit_dialog_
message), Toast.LENGTH_SHORT).show();
            tExit=new Timer();
            tExit.schedule(new TimerTask() {
                public void run() {
                    isExit=false;
                }
            }, 2000);
        } else {
            MainApplication.getInstance().exit();
        }
    }
}
```

7.4.3 情景模式页面

实现“返回”和六种“情景模式”。“返回”功能的作用是返回上一级页面；当点击“起床模式”时，系统把“起床模式”唯一的 ID 号经网络传输到智能网关上，智能网关接收到模式 ID 号，与智能网关的数据库中的模式 ID 进行比对，读取与模式相关的控制信息来实现对相关设备的控制。“离家模式”“回家模式”“就餐模式”“影院模式”“睡眠模式”与“起床模式”的实现类似，只是模式 ID 和相关的控制信息不同而已；情景模式页面如图 7-6 所示。

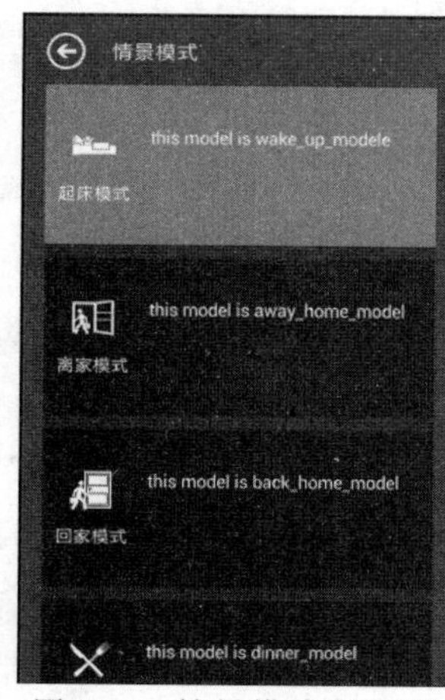

图 7-6 情景模式页面图

程序的界面布局代码如下：CHOME\res\layout\activity_model_list.xml。

```
<?xml version="1.0" encoding="utf-8"?>
<RelativeLayout xmlns:android="http://schemas. android.
com/apk/res/android"
    android:layout_width="fill_parent"
    android:layout_height="fill_parent"
    android:background="@drawable/main_default_bg" >
    <include
        android:id="@+id/bt_createtask_title_layout"
        layout="@layout/activity_common_title_bar" />
    <RelativeLayout
        android:id="@+id/RelativeLayout1"
        android:layout_width="match_parent"
```

```
        android:layout_height="match_parent"
        android:layout_below="@id/bt_createtask_title_layout"
        android:layout_centerHorizontal="true"
        android:layout_marginLeft="15dp"
        android:layout_marginRight="15dp"
        android:background="@drawable/content_default_bg"
        android:orientation="vertical"
        android:visibility="visible" >
        <GridView
            android:id="@+id/gdv_models"
            android:layout_width="match_parent"
            android:layout_height="wrap_content"
            android:layout_alignParentLeft="true"
            android:layout_alignParentTop="true"
            android:horizontalSpacing="10dp"
            android:listSelector="@drawable/selector_list"
            android:numColumns="1"
            android:verticalSpacing="10dp"
            android:visibility="visible" >
        </GridView>
        <include
            android:id="@+id/view_loading_error"
            android:layout_width="match_parent"
            android:layout_height="match_parent"
            android:layout_alignParentLeft="true"
            android:layout_below="@+id/gdv_models"
            layout="@layout/view_loading_error"
            android:visibility="gone" />
    </RelativeLayout>
</RelativeLayout>
```

程序清单：CHOME\src\com\zigcloud\smarthome\activity\ModelListActivity.java。

```
public class ModelListActivity extends BaseActivity{/*** 情景模式**/
    private TextView titlebar_left;
    private TextView titlebar_title;
    private GridView modelsGridView;
    private ArrayList<HashMap<String,Object>> modelsArrayList=new ArrayList
<HashMap<String, Object>>();
    private HashMap<String,Object> modelsHashMap=new HashMap<String, Object>();
    private RadioAdapter modelsAdapter;
    /*** 获取模式列表**/
    private    AsyncTask<String,Integer,ModelListHttpRequestTask.Result>
modelListHttpRequestTask=new ModelListHttpRequestTask();
    /*** 切换当前模式**/
    private       AsyncTask<String,Integer,ControlResultJson>changeModel
HttpRequestTask=new ChangeModelHttpRequestTask();
    public void onCreate(Bundle savedInstanceState) {
        super.onCreate(savedInstanceState);
        setContentView(R.layout.activity_model_list);
        MainApplication.getInstance().addActivity(this);
        titlebar_left=(TextView) findViewById(R.id.titlebar_left);
        titlebar_left.setOnClickListener(new View.OnClickListener() {
            public void onClick(View v) {
                onBackPressed();
            }
```

```
        });
        titlebar_title=(TextView) findViewById(R.id.titlebar_title);
        titlebar_title.setText(R.string.title_models);
        initialModelsGridView();
    }
    private void initialModelsGridView(){/*** 初始化情景模式列表**/
        modelsGridView=(GridView) findViewById(R.id.gdv_models);
        modelsAdapter=new RadioAdapter(getApplicationContext(),
            modelsArrayList, R.layout.activity_model_list_item,
            new String[] {"ItemImage","ItemText","ItemContent","ItemId"},
            new int[] {R.id.itemImage,R.id.itemText,R.id.itemContent},
            R.drawable.metro_home_blocks,
            R.drawable.metro_home_blocks_hover);
        modelsGridView.setAdapter(modelsAdapter);
        modelsGridView.setOnItemClickListener(new OnItemClickListener() {
            @SuppressWarnings("unchecked")
            public void onItemClick(AdapterView<?> arg0,View arg1,int arg2,
long arg3) {
                HashMap<String,Object>item=(HashMap<String,Object>) arg0.
getItemAtPosition(arg2);
                if(item!=null&&item.get("ItemId")!=null&&item.get("ItemId").
toString()!=null){ //改变模式
                if(changeModelHttpRequestTask.getStatus()!=Status.RUNNING){
                    changeModelHttpRequestTask=new ChangeModelHttpRequestTask() ;
                    changeModelHttpRequestTask.execute(String.valueOf(arg2),
item.get("ItemId").toString());
                        modelsAdapter.setSeclectItemPosition(arg2);
                        modelsAdapter.notifyDataSetChanged();
                    }
                }
            }
        });
        modelListHttpRequestTask.execute();
    }
    private ModelDAO modelDAO=new ModelDAO();
    /*** 获取模式列表**/
    private class ModelListHttpRequestTask extends AsyncTask<String, Integer,
ModelListHttpRequestTask.Result>{
        protected ModelListHttpRequestTask.Result doInBackground(String... params){
            ModelJson currenModelEntity=modelDAO.getCurrentModel();
            String currentModelId=currenModelEntity==null?null:currenModelEntity.id;
            return new ModelListHttpRequestTask.Result(currentModelId,modelDAO. getAll());
        }
        protected void onPostExecute(ModelListHttpRequestTask.Result result) {
            if(result!=null&& result.getItems()!=null){
                ModelJson modelEntity=null;
```

```
            for(int i=0;i<result.getItems().size();i++){
                modelEntity= result.getItems().get(i);
                if(modelEntity!=null){
                    modelsHashMap=new HashMap<String, Object>();
                    modelsHashMap.put("ItemImage",R.drawable.metro_home_ blacks_
scan_code);
                    modelsHashMap.put("ItemText", modelEntity.name);
                    modelsHashMap.put("ItemContent", modelEntity.description);
                    modelsHashMap.put("ItemId", modelEntity.id);
                    modelsArrayList.add(modelsHashMap);
                    if(modelEntity.id.equals(result.selectedId)){
                        modelsAdapter.setSeclectItemPosition(i);
                    }
                }
            }
            modelsAdapter.notifyDataSetChanged();
        }
        else{
            findViewById(R.id.view_loading_error).setVisibility(0);
        }
    }
    public class Result{/*** 获取模式的返回结果**/
        public String  selectedId; /*** 当前选中的模式**/
        public List<ModelJson> items; /*** 模式列表**/
        public Result(String selectedId,List<ModelJson> items){
            this.selectedId=selectedId;
            this.items=items;
        }
        @SuppressWarnings("unused")
        public String getSelectedId() {
            return selectedId;
        }
        @SuppressWarnings("unused")
        public void setSelectedId(String selectedId) {
            this.selectedId=selectedId;
        }
        public List<ModelJson> getItems() {
            return items;
        }
        @SuppressWarnings("unused")
        public void setItems(List<ModelJson> items) {
            this.items=items;
        }
    }
}
```

```
    private class ChangeModelHttpRequestTask extends AsyncTask<String, Integer,
ControlResultJson>{/*** 切换当前模式 **/
        protected ControlResultJson doInBackground(String... params){
            if(params!=null&&params.length>1){
                return modelDAO.changeModelById(params[1]);
            }
            return null;
        }
        protected void onPostExecute(ControlResultJson result) {
            modelsAdapter.notifyDataSetInvalidated();
        }
    }
}
```

7.4.4 我的房间页面

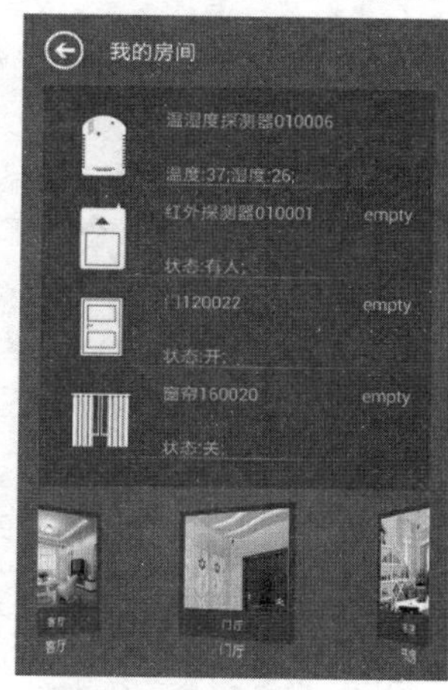

图 7-7　我的房间页面

实现“返回”和“我的房间”设备列表功能，页面布局如图 7-7 所示。

“返回”功能的作用是返回上一级页面；用户可以通过点击设备列表中的设备，进入具体设备的控制页面中。

程序的界面布局代码如下：CH07\CHOME\res\layout\ activity_room_list.xml。

```
<?xml version="1.0" encoding="utf-8"?>
<RelativeLayout xmlns:android="http://schemas.android.
com/apk/res/android"
    android:layout_width="fill_parent"
    android:layout_height="fill_parent"
    android:background="@drawable/main_default_bg" >
    <include
        android:id="@id/bt_createtask_title_layout"
        layout="@layout/activity_common_title_bar" />
    <RelativeLayout
        android:id="@+id/RelativeLayout1"
        android:layout_width="match_parent"
        android:layout_height="match_parent"
        android:layout_below="@id/bt_createtask_title_layout"
        android:layout_centerHorizontal="true"
        android:layout_marginLeft="15dp"
        android:layout_marginRight="15dp"
        android:background="@drawable/content_default_bg"
        android:orientation="vertical"
        android:visibility="visible" >
        <com.zigcloud.smarthome.widget.Gallery3D
            android:id="@+id/gal_rooms"
            android:layout_width="fill_parent"
            android:layout_height="wrap_content"
```

```
        android:layout_alignParentBottom="true"
        android:layout_alignParentLeft="true"
        android:spacing="30dp"
        android:unselectedAlpha="128" />
    <LinearLayout
        android:id="@+id/LinearLayout1"
        android:layout_width="match_parent"
        android:layout_height="match_parent"
        android:layout_above="@+id/gal_rooms"
        android:layout_alignParentTop="true"
        android:orientation="vertical" >
        <include
            android:id="@+id/view_loading_error"
            android:layout_width="match_parent"
            android:layout_height="match_parent"
            layout="@layout/view_loading_error"
            android:visibility="gone" />
        <include
            android:id="@+id/view_loading"
            android:layout_width="match_parent"
            android:layout_height="match_parent"
            layout="@layout/view_loading"
            android:visibility="gone" />
        <GridView
            android:id="@+id/gdv_equipments"
            android:layout_width="match_parent"
            android:layout_height="wrap_content"
            android:numColumns="1"
            android:padding="10dp"
            android:verticalSpacing="10dp"
            android:visibility="visible" >
        </GridView>
    </LinearLayout>
  </RelativeLayout>
</RelativeLayout>
```

程序清单：CH07\CHOME\src\com\zigcloud\smarthome\activity\RoomListListActivity.java。

```
@SuppressLint({ "CutPasteId", "HandlerLeak" })
public class RoomListActivity extends BaseActivity{ /*** 我的房间**/
  private TextView titlebar_left;
  private TextView titlebar_title;
  private GridView equipmentsGridView;
  private ArrayList<HashMap<String, Object>> equipmentsArrayList=new ArrayList
<HashMap<String, Object>>();
  private HashMap<String, Object> equipmentsHashMap = new HashMap<String, Object>();
  private SimpleAdapter equipmentsAdapter;
  private EquipmentListHttpRequestTask equipmentListHttpRequestTask=new
EquipmentListHttpRequestTask();/*** 获取设备列表**/
  public void onCreate(Bundle savedInstanceState) {
    super.onCreate(savedInstanceState);
    setContentView(R.layout.activity_room_list);
    MainApplication.getInstance().addActivity(this);
    titlebar_left=(TextView) findViewById(R.id.titlebar_left);
    titlebar_left.setOnClickListener(new View.OnClickListener() {
      public void onClick(View v) {
```

```
                onBackPressed();
            }
        });
        titlebar_title=(TextView) findViewById(R.id.titlebar_title);
        titlebar_title.setText(R.string.title_myroom);
        initialRoomsGallery();
        initialEquipmentsGridView();
    }
    private void initialEquipmentsGridView(){/*** 初始化设备信息列表 **/
        equipmentsGridView=(GridView) findViewById(R.id.gdv_equipments);
        equipmentsAdapter=new SimpleAdapter(getApplicationContext(),
equipmentsArrayList,R.layout.activity_equipment_list_item,new String[] {"ItemImage",
"ItemNodeName","ItemDataValue", "ItemArea"},new int[] {R.id.img_image,R.id.
tv_name,R.id.tv_datavalue, R.id.tv_area});
        equipmentsGridView.setAdapter(equipmentsAdapter);
        equipmentsGridView.setOnItemClickListener(new OnItemClickListener() {
            @SuppressWarnings("unchecked")
            public void onItemClick(AdapterView<?>arg0,View arg1,int arg2,long arg3){
                HashMap<String,Object>item=(HashMap<String,Object>) arg0.
getItemAtPosition(arg2);
                String nodeId=item.get("ItemNodeId")==null ?null:item.get
("ItemNodeId").toString();
                String nodeName=item.get("ItemNodeName")==null ?null:item.get
("ItemNodeName").toString();
                String nodeTypeId=item.get("ItemNodeTypeId")==null ?null:item.
get("ItemNodeTypeId").toString();
                String nodeTypeName=item.get("ItemNodeTypeName")==null ?null:
item.get("ItemNodeTypeName").toString();
                String dataValueString=item.get("ItemDataValueString")==null ?
null:item.get("ItemDataValueString").toString();
                String updateTimeString=item.get("ItemUpdateTimeString")== null
?null:item.get("ItemUpdateTimeString").toString();
                Intent intent =new Intent(getApplicationContext(),
EquipmentActivity. class);
                intent.putExtra("nodeId", nodeId);
                intent.putExtra("nodeName", nodeName);
                intent.putExtra("nodeTypeId", nodeTypeId);
                intent.putExtra("nodeTypeName", nodeTypeName);
                intent.putExtra("dataValueString", dataValueString);
                intent.putExtra("updateTimeString",updateTimeString);
                startActivity(intent);
            }
        });
        equipmentListHttpRequestTask.execute();
    }
    private EquipmentDAO mEquipmentDAO=new EquipmentDAO();
    private Handler mHandler=new Handler(){
        public void handleMessage(Message msg) {
            super.handleMessage(msg);
            switch(msg.what){
                case 10001: //正在加载
                    findViewById(R.id.gdv_equipments).setVisibility(8);
                    findViewById(R.id.view_loading).setVisibility(0);
                    findViewById(R.id.view_loading_error).setVisibility(8);
                    break;
```

```
            case 10002: //加载成功
               findViewById(R.id.gdv_equipments).setVisibility(0);
               findViewById(R.id.view_loading).setVisibility(8);
               findViewById(R.id.view_loading_error).setVisibility(8);
               break;
            case 10003: //加载失败
               findViewById(R.id.gdv_equipments).setVisibility(8);
               findViewById(R.id.view_loading).setVisibility(8);
               findViewById(R.id.view_loading_error).setVisibility(0);
               break;
         }
      }
   };
   private class EquipmentListHttpRequestTask extends AsyncTask<String,
Integer, List<EquipmentJson>>{/*** 获取 equipment 列表**/
      protected void onPreExecute() {
         super.onPreExecute();
         if(!(equipmentsAdapter!=null&&equipmentsAdapter.getCount()>0)){
            mHandler.sendEmptyMessage(10001);
         }
      }
      protected List<EquipmentJson> doInBackground(String... params) {
         return mEquipmentDAO.getAll();
      }
      protected void onProgressUpdate(Integer... values) {
         super.onProgressUpdate(values);
      }
      protected void onPostExecute(List<EquipmentJson> result) {
         if(result!=null){
            EquipmentJson equipmentEntity=null;
            BaseEquipmentEntity equipment=null;
            for(int i=0;i<result.size();i++){
               equipmentEntity= result.get(i);
               equipment= BaseEquipmentEntity.parse(equipmentEntity);
               if(equipment!=null){
                  equipmentsHashMap=new HashMap<String, Object>();
                  equipmentsHashMap.put("ItemImage", equipment.getIconRes());
                  equipmentsHashMap.put("ItemNodeName", equipment.getName());
                  equipmentsHashMap.put("ItemDataValue",
equipment.getDataValueString());
                  equipmentsHashMap.put("ItemArea", "未知区域");
                  equipmentsHashMap.put("ItemNodeId",equipment. getNodeId());
                  equipmentsHashMap.put("ItemNodeTypeId",equipment.getNode
TypeId());
                  equipmentsHashMap.put("ItemNodeTypeName",equipment.
getTypeName());
                  equipmentsHashMap.put("ItemDataValueString",equipment.
getDataValue String());
                  equipmentsHashMap.put("ItemUpdateTimeString",equipment.
getUpdateTimeString());
                  equipmentsArrayList.add(equipmentsHashMap);
               }
            }
            equipmentsAdapter.notifyDataSetChanged();
            mHandler.sendEmptyMessage(10002);
```

```
            }
            else{
               mHandler.sendEmptyMessage(10003);
            }
         }
      };
      private void initialRoomsGallery(){
         Gallery3D gallery = (Gallery3D) findViewById(R.id.gal_rooms);
         ArrayList<HashMap<String,Object>>lstImageItem=new ArrayList<HashMap
<String, Object>>();
         HashMap<String, Object> map=new HashMap<String, Object>();
         map.put("ItemImage", R.drawable.scene_gallery_0);
         map.put("ItemText", "大厅");
         map.put("ItemContent", "打开家里的灯光，关闭安防设备");
         lstImageItem.add(map);
         map=new HashMap<String, Object>();
         map.put("ItemImage", R.drawable.scene_gallery_1);
         map.put("ItemText", "书房");
         map.put("ItemContent", "打开家里的灯光，关闭安防设备");
         lstImageItem.add(map);
         map=new HashMap<String, Object>();
         map.put("ItemImage", R.drawable.scene_gallery_2);
         map.put("ItemText", "卧室");
         map.put("ItemContent", "打开家里的灯光，关闭安防设备");
         lstImageItem.add(map);
         map=new HashMap<String, Object>();
         map.put("ItemImage", R.drawable.scene_gallery_3);
         map.put("ItemText", "阳台");
         map.put("ItemContent", "打开家里的灯光，关闭安防设备");
         lstImageItem.add(map);
         SimpleAdapter saImageItems=new SimpleAdapter(this,lstImageItem,R.
layout.view_grallery3d_item,new String[]{"ItemImage", "ItemText","ItemContent"},
new int[] {R.id.itemImage,R.id.itemText,R.id.itemContent});
         gallery.setFadingEdgeLength(0);
         gallery.setAdapter(saImageItems);
         gallery.setOnItemSelectedListener(new OnItemSelectedListener() {
            public void onItemSelected(AdapterView<?> parent, View view, int
position, long id) {
         Toast.makeText(RoomListActivity.this, "img " + (position+1) + " selected",
Toast.LENGTH_SHORT).show();
            }
            public void onNothingSelected(AdapterView<?> parent) {
            }
         });
         gallery.setOnItemClickListener(new OnItemClickListener() {
            public void onItemClick(AdapterView<?> parent, View view, int
position, long id) {
               Toast.makeText(RoomListActivity.this, "img " + (position+1) +
"selected", Toast.LENGTH_SHORT).show();
            }
         });
      }
   }
```

7.4.5 具体设备页面

实现“返回”设备名称显示、设备图片呈现、“开启”“关闭”功能。页面布局如图 7-8 所示。

图 7-8 设备页面控制图

“返回”功能的作用是返回上一级页面；在“返回”后面显示当前设备的名称。设备图片采用设备实物图；“开”按键和“关”按键分别实现对设备的开关控制。程序的界面布局代码如下：

程序清单：CH07\CHOME\res\layout\activity_room_list.xml。

```
<?xml version="1.0" encoding="utf-8"?>
<RelativeLayout
xmlns:android="http://schemas.android.com/apk/
res/android"
    android:layout_width="fill_parent"
    android:layout_height="fill_parent"
    android:background="@drawable/main_default_bg" >
    <include
        android:id="@id/bt_createtask_title_ layout"
        layout="@layout/activity_common_title_ bar" />
    <RelativeLayout
        android:id="@+id/RelativeLayout1"
        android:layout_width="match_parent"
        android:layout_height="match_parent"
        android:layout_below="@id/bt_createtask_ title_layout"
        android:layout_centerHorizontal="true"
        android:layout_marginLeft="15dp"
        android:layout_marginRight="15dp"
        android:background="@drawable/content_ default_bg"
        android:orientation="vertical"
        android:visibility="visible" >
        <com.zigcloud.smarthome.widget.Gallery3D
            android:id="@+id/gal_rooms"
            android:layout_width="fill_parent"
            android:layout_height="wrap_content"
            android:layout_alignParentBottom="true"
            android:layout_alignParentLeft="true"
            android:spacing="30dp"
            android:unselectedAlpha="128" />
        <LinearLayout
            android:id="@+id/LinearLayout1"
            android:layout_width="match_parent"
            android:layout_height="match_parent"
            android:layout_above="@+id/gal_rooms"
            android:layout_alignParentTop="true"
```

```
            android:orientation="vertical" >
            <include
                android:id="@+id/view_loading_error"
                android:layout_width="match_parent"
                android:layout_height="match_parent"
                layout="@layout/view_loading_error"
                android:visibility="gone" />
            <include
                android:id="@+id/view_loading"
                android:layout_width="match_parent"
                android:layout_height="match_parent"
                layout="@layout/view_loading"
                android:visibility="gone" />
            <LinearLayout
               android:id="@+id/ll_equipment"
               android:layout_width="match_parent"
               android:layout_height="match_parent"
               android:orientation="vertical" >
            </LinearLayout>
        </LinearLayout>
    </RelativeLayout>
</RelativeLayout>
```

程序清单：CH07\CHOME\src\com\zigcloud\smarthome\activity\EquipmentActivity.java。

```
public class EquipmentActivity extends BaseActivity{ /*** 设备* */
   private TextView titlebar_left;
   private TextView titlebar_title;
   private String nodeId;
   private String nodeName;
   private String nodeTypeId;
   private String nodeTypeName;
   private  String dataValueString;
   private String updateTimeString;
   protected void onCreate(Bundle savedInstanceState) {
      super.onCreate(savedInstanceState);
      setContentView(R.layout.activity_equipment);
      MainApplication.getInstance().addActivity(this);
      nodeId=getIntent().getStringExtra("nodeId");
      nodeName=getIntent().getStringExtra("nodeName");
      nodeTypeId=getIntent().getStringExtra("nodeTypeId");
      nodeTypeName=getIntent().getStringExtra("nodeTypeName");
      dataValueString=getIntent().getStringExtra("dataValueString");
      updateTimeString=getIntent().getStringExtra("updateTimeString");
      titlebar_left=(TextView) findViewById(R.id.titlebar_left);
      titlebar_left.setOnClickListener(new View.OnClickListener() {
         public void onClick(View v) {
            onBackPressed();
```

```
            }
        });
        titlebar_title=(TextView) findViewById(R.id.titlebar_title);
        titlebar_title.setText(nodeName);
        initalEquipment();
    }
    private void initalEquipment(){/*** 初始化设备信息 **/
        LinearLayout ll_equipment=(LinearLayout)findViewById(R.id.ll_ equipment);
        if(ll_equipment!=null){
            if(nodeTypeId!=null){
                BaseEquipmentWidget equipmentWidget=null;
                if(nodeTypeId.equals("20")){
                    equipmentWidget=new CurtainWidget(EquipmentActivity.this);
                }
                else if(nodeTypeId.equals("24")){
                    equipmentWidget=new LampWidget(EquipmentActivity.this);
                }
                else if(nodeTypeId.equals("25")){
                    equipmentWidget=new BedLampWidget(EquipmentActivity.this);
                }
                else if(nodeTypeId.equals("26")){
                    equipmentWidget=new WallLampWidget(EquipmentActivity.this);
                }
                else{
                    equipmentWidget=new BaseEquipmentWidget(EquipmentActivity.this);
                }
                equipmentWidget.setNodeId(nodeId);
                equipmentWidget.setNodeTypeName(nodeTypeName);
                equipmentWidget.setNodeDataString(dataValueString);
                equipmentWidget.setUpdateTimeString(updateTimeString);
                equipmentWidget.addSendCmdListener(new BaseEquipmentWidget.
Listener(){
                    public void sendCmd(String nodeId, String stateFlag,String
stateValue) {
                        if(mSendCmdHttpRequestTask.getStatus()!=Status.RUNNING){
                            mSendCmdHttpRequestTask=new SendCmdHttpRequestTask();
                            mSendCmdHttpRequestTask.execute(nodeId,stateFlag, stateValue);
                        }
                    }
                });
                ll_equipment.addView(equipmentWidget,new
LayoutParams(LayoutParams.MATCH_PARENT,LayoutParams.MATCH_PARENT));
            }
        }
    }
    protected EquipmentDAO mEquipmentDAO=new EquipmentDAO();
```

```
    /*** 设备控制业务类**/
    /***异步发送命令任务类**/
    protected SendCmdHttpRequestTask mSendCmdHttpRequestTask=new SendCmd
HttpRequestTask();
    /***切换当前模式**/
    public class SendCmdHttpRequestTask extends  AsyncTask<String, Integer,
ControlResultJson>{
        protected ControlResultJson doInBackground(String... params) {
            if(params!=null&&params.length>2){
                mEquipmentDAO.sendCmd(params[0], params[1], params[2]);
            }
            return null;
        }
    }
    protected EquipmentHttpRequestTask mEquipmentHttpRequestTask=new
EquipmentHttpRequestTask();
    /*** 获取 equipment**/
    protected  class  EquipmentHttpRequestTask  extends  AsyncTask<String,
Integer, EquipmentJson>{
        protected void onPreExecute() {
            super.onPreExecute();
        }
        protected void onProgressUpdate(Integer... values) {
            super.onProgressUpdate(values);
        }
        protected EquipmentJson doInBackground(String... arg0) {
            String nodeId=arg0!=null&&arg0.length>0?arg0[0]:null;
            return mEquipmentDAO.getByNodeId(nodeId);
        }
        protected void onPostExecute(EquipmentJson result) {
            super.onPostExecute(result);
            BaseEquipmentEntity baseEquipment=BaseEquipmentEntity.parse(result);
            if(baseEquipment!=null){
            }
        }
    }
}
```

7.4.6 家用电器页面

实现“返回”和六个示图功能。页面布局如图 7-9 所示。

“返回”功能的作用是返回上一级页面；用户可以通过点击设备列表中的设备，进入具体设备的监控界面。

程序的界面布局代码如下：CH07\CHOME\res\layout\activity_function_list.xml。

```
<?xml version="1.0" encoding="utf-8"?>
<RelativeLayout xmlns:android="http://schemas.
android.com/apk/res/android"
```

图 7-9 家用电器页面列表图

```
    android:layout_width="fill_parent"
    android:layout_height="fill_parent"
    android:background="@drawable/main_default_bg" >
    <include
        android:id="@+id/bt_createtask_title_layout"
        layout="@layout/activity_common_title_bar" />
    <RelativeLayout
        android:id="@+id/RelativeLayout1"
        android:layout_width="match_parent"
        android:layout_height="match_parent"
        android:layout_below="@id/bt_createtask_title_layout"
        android:layout_centerHorizontal="true"
        android:layout_marginLeft="15dp"
        android:layout_marginRight="15dp"
        android:background="@drawable/content_default_bg"
        android:orientation="vertical"
        android:visibility="visible" >
        <GridView
            android:id="@+id/gdv_models"
            android:layout_width="match_parent"
            android:layout_height="wrap_content"
            android:layout_alignParentLeft="true"
            android:layout_alignParentTop="true"
            android:horizontalSpacing="10dp"
            android:listSelector="@drawable/selector_list"
            android:numColumns="1"
            android:verticalSpacing="10dp"
            android:visibility="visible" >
        </GridView>
        <include
            android:id="@+id/view_loading_error"
            android:layout_width="match_parent"
            android:layout_height="match_parent"
            android:layout_alignParentLeft="true"
            android:layout_below="@+id/gdv_models"
            layout="@layout/view_loading_error"
            android:visibility="gone" />
    </RelativeLayout>
</RelativeLayout>
```

程序清单：CH07\CHOME\src\com\zigcloud\smarthome\activity\FunctionListActivity.java。

```
package com.zigcloud.smarthome.activity;
@SuppressLint({ "CutPasteId", "HandlerLeak" })
public class FunctionListActivity extends BaseActivity{/*** 家用电器 **/
    private TextView titlebar_left;
    private TextView titlebar_title;
    private GridView equipmentsGridView;
```

```
    private ArrayList<HashMap<String, Object>> equipmentsArrayList=new ArrayList
<HashMap<String, Object>>();
    private HashMap<String,Object> equipmentsHashMap=new HashMap<String,Object>();
    private SimpleAdapter equipmentsAdapter;
    /*** 获取设备列表**/
    private EquipmentListHttpRequestTask equipmentListHttpRequestTask=new
EquipmentListHttpRequestTask();
    public void onCreate(Bundle savedInstanceState) {
       super.onCreate(savedInstanceState);
       setContentView(R.layout.activity_room_list);
       MainApplication.getInstance().addActivity(this);
       titlebar_left=(TextView) findViewById(R.id.titlebar_left);
       titlebar_left.setOnClickListener(new View.OnClickListener() {
          public void onClick(View v) {
             onBackPressed();
          }
       });
       titlebar_title=(TextView) findViewById(R.id.titlebar_title);
       titlebar_title.setText(R.string.title_myfunction);
       initialRoomsGallery();
       initialEquipmentsGridView();
    }
    private void initialEquipmentsGridView(){/*** 初始化设备信息列表**/
       equipmentsGridView=(GridView) findViewById(R.id.gdv_equipments);
       equipmentsAdapter=new SimpleAdapter(getApplicationContext(),
equipmentsArrayList,R.layout.activity_equipment_list_item,new String[] {"ItemImage",
"ItemNodeName","ItemDataValue", "ItemArea"},new int[] {R.id.img_image,R.id.
tv_name,R.id.tv_datavalue, R.id.tv_area});
       equipmentsGridView.setAdapter(equipmentsAdapter);
       equipmentsGridView.setOnItemClickListener(new OnItemClickListener(){
          @SuppressWarnings("unchecked")
          public void onItemClick(AdapterView<?> arg0, View arg1, int arg2,
long arg3) {
             HashMap<String, Object> item=( HashMap<String, Object>) arg0.
getItemAtPosition(arg2);
             String nodeId=item.get("ItemNodeId")==null ?null:item.get
("ItemNodeId").toString();
             String nodeName=item.get("ItemNodeName")==null ?null:item. get
("ItemNodeName").toString();
             String nodeTypeId=item.get("ItemNodeTypeId")==null ?null: item.get
("ItemNodeTypeId").toString();
             String nodeTypeName=item.get("ItemNodeTypeName")==null ?null:
item.get("ItemNodeTypeName").toString();
             String dataValueString=item.get("ItemDataValueString")== null ?
null:item.get("ItemDataValueString").toString();
             String  updateTimeString=item.get("ItemUpdateTimeString")==n
ull ?null:item.get("ItemUpdateTimeString").toString();
             Intent intent =new Intent(getApplicationContext(),EquipmentActivity.
class);
             intent.putExtra("nodeId", nodeId);
```

```
            intent.putExtra("nodeName", nodeName);
            intent.putExtra("nodeTypeId", nodeTypeId);
            intent.putExtra("nodeTypeName", nodeTypeName);
            intent.putExtra("dataValueString", dataValueString);
            intent.putExtra("updateTimeString",updateTimeString);
            startActivity(intent);
         }
      });
      equipmentListHttpRequestTask.execute();
   }
   private EquipmentDAO mEquipmentDAO=new EquipmentDAO();
   private Handler mHandler=new Handler(){
      public void handleMessage(Message msg) {
         super.handleMessage(msg);
         switch(msg.what){
            case 10001:               //正在加载
               findViewById(R.id.gdv_equipments).setVisibility(8);
               findViewById(R.id.view_loading).setVisibility(0);
               findViewById(R.id.view_loading_error).setVisibility(8);
               break;
            case 10002:               //加载成功
               findViewById(R.id.gdv_equipments).setVisibility(0);
               findViewById(R.id.view_loading).setVisibility(8);
               findViewById(R.id.view_loading_error).setVisibility(8);
               break;
            case 10003:               //加载失败
               findViewById(R.id.gdv_equipments).setVisibility(8);
               findViewById(R.id.view_loading).setVisibility(8);
               findViewById(R.id.view_loading_error).setVisibility(0);
               break;
         }
      }
   };
   private class EquipmentListHttpRequestTask extends AsyncTask<String,
Integer, List<EquipmentJson>>{/*** 获取 equipment 列表**/
      protected void onPreExecute() {
         super.onPreExecute();
         if(!(equipmentsAdapter!=null&&equipmentsAdapter.getCount()>0)){
            mHandler.sendEmptyMessage(10001);
         }
      }
      protected List<EquipmentJson> doInBackground(String... params) {
         return mEquipmentDAO.getAll();
      }
      protected void onProgressUpdate(Integer... values) {
         super.onProgressUpdate(values);
      }
      protected void onPostExecute(List<EquipmentJson> result) {
         if(result!=null){
            EquipmentJson equipmentEntity=null;
```

```
            BaseEquipmentEntity equipment=null;
            for(int i=0;i<result.size();i++){
                equipmentEntity=result.get(i);
                equipment=BaseEquipmentEntity.parse(equipmentEntity);
                if(equipment!=null){
                    equipmentsHashMap=new HashMap<String, Object>();
                    equipmentsHashMap.put("ItemImage", equipment.getIconRes());
                    equipmentsHashMap.put("ItemNodeName", equipment.getName());
                    equipmentsHashMap.put("ItemDataValue",
equipment.getDataValueString());
                    equipmentsHashMap.put("ItemArea", "未知区域");
                    equipmentsHashMap.put("ItemNodeId",equipment.getNodeId());
                    equipmentsHashMap.put("ItemNodeTypeId",equipment.
getNodeTypeId());
                    equipmentsHashMap.put("ItemNodeTypeName",equipment.
getTypeName());
                    equipmentsHashMap.put("ItemDataValueString",equipment.
getDataValueString());
                    equipmentsHashMap.put("ItemUpdateTimeString",equipment.
getUpdateTimeString());
                    equipmentsArrayList.add(equipmentsHashMap);
                }
            }
            equipmentsAdapter.notifyDataSetChanged();
            mHandler.sendEmptyMessage(10002);
        }
        else{
            mHandler.sendEmptyMessage(10003);
        }
    }
};
private void initialRoomsGallery(){
    Gallery3D gallery=(Gallery3D) findViewById(R.id.gal_rooms);
    ArrayList<HashMap<String,Object>> lstImageItem=new ArrayList <HashMap
<String, Object>>();
    HashMap<String, Object> map=new HashMap<String, Object>();
    map.put("ItemImage", R.drawable.scene_gallery_0);
    map.put("ItemText", "环境");
    map.put("ItemContent", "打开家里的灯光，关闭安防设备");
    lstImageItem.add(map);
    map=new HashMap<String, Object>();
    map.put("ItemImage", R.drawable.scene_gallery_1);
    map.put("ItemText", "灯");
    map.put("ItemContent", "打开家里的灯光，关闭安防设备");
    lstImageItem.add(map);
    map=new HashMap<String, Object>();
    map.put("ItemImage", R.drawable.scene_gallery_2);
    map.put("ItemText", "电视机");
    map.put("ItemContent", "打开家里的灯光，关闭安防设备");
    lstImageItem.add(map);
```

```
        map=new HashMap<String, Object>();
        map.put("ItemImage", R.drawable.scene_gallery_3);
        map.put("ItemText", "窗帘");
        map.put("ItemContent", "打开家里的灯光，关闭安防设备");
        lstImageItem.add(map);
        SimpleAdapter saImageItems=new SimpleAdapter(this,lstImageItem,R.
layout.view_grallery3d_item,new String[] {"ItemImage", "ItemText","ItemContent"},
new int[] {R.id.itemImage,R.id. itemText,R.id.itemContent});

        gallery.setFadingEdgeLength(0);
        gallery.setAdapter(saImageItems);
        gallery.setOnItemSelectedListener(new OnItemSelectedListener() {

            public void onItemSelected(AdapterView<?> parent, View view, int
position, long id) {
                Toast.makeText(RoomListActivity.this, "img " + (position+1) +
" selected",Toast.LENGTH_SHORT).show();
            }
            public void onNothingSelected(AdapterView<?> parent) {
            }
        });
        gallery.setOnItemClickListener(new OnItemClickListener() {

            public void onItemClick(AdapterView<?> parent, View view, int
position, long id) {
                Toast.makeText(RoomListActivity.this, "img " + (position+1) +
" selected", Toast.LENGTH_SHORT).show();
            }
        });
    }
}
```

具体代码实现请见光盘中的系统源代码。

7.5 本章小结

本章通过智能家居项目，让学生了解企业项目的整体开发流程：项目立项、需求分析、总体设计、详细设计、编码实现、系统调试。同时通过这个项目理解以前章节中学习到的知识并灵活应用到实现项目中。

第8章 Android 校园通项目

学习目标：

- DAO 设计模式；
- XML 解析；
- 服务器数据访问；
- 异步任务等技术。

8.1 项目概述

“校园通”软件是一款主要服务于大一新生的软件。在大一新生没有进入大学校园之前，对大学充满了好奇，并且对新的环境充满陌生感。这款软件针对新生入学时想要知道的各种信息资讯做了分类，并且还能对之后的大学生活和学习有较大的帮助。资讯分类如图 8-1 所示。

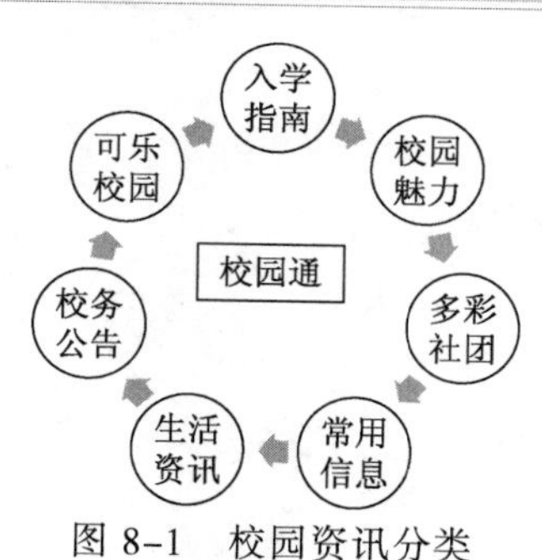

图 8-1　校园资讯分类

8.2 项目设计

8.2.1　项目总体功能需求

项目总体需求包括 Android 客户端子系统和后台服务器子系统，如表 8-1 所示。

表 8-1　校园通项目需求

子系统	需求编号	需求名称	说明
Android 客户端	IFA-0010	Android 应用框架	在前面已经实现，有详细的需求描述
Android 客户端	IFA-0020	资讯分类展示	可导航展示各类资讯的内容，包括文字、图片等
Android 客户端	IFA-0030	资讯详细信息	通过资讯分类展示，导航到某个资讯后，可以查看其详细信息，包括标题、内容以及相关图片等
Android 客户端	IFA-0040	资讯更新	可以从服务器更新最新的资讯内容到手机
后台服务器	IFW-0010	资讯平台管理	主要对整个项目的一些全局属性进行设置，如项目名称、创建人、创建时间等信息

续表

子 系 统	需 求 编 号	需 求 名 称	说 明
后台服务器	IFW-0020	资讯分类管理	实现对资讯分类的增加、删除、修改等操作
后台服务器	IFW-0030	资讯管理	实现对资讯信息的增加、删除、修改等操作，包括标题、内容以及相关图片等

8.2.2 项目总体设计

整个项目物理架构设计如图 8-2 所示。

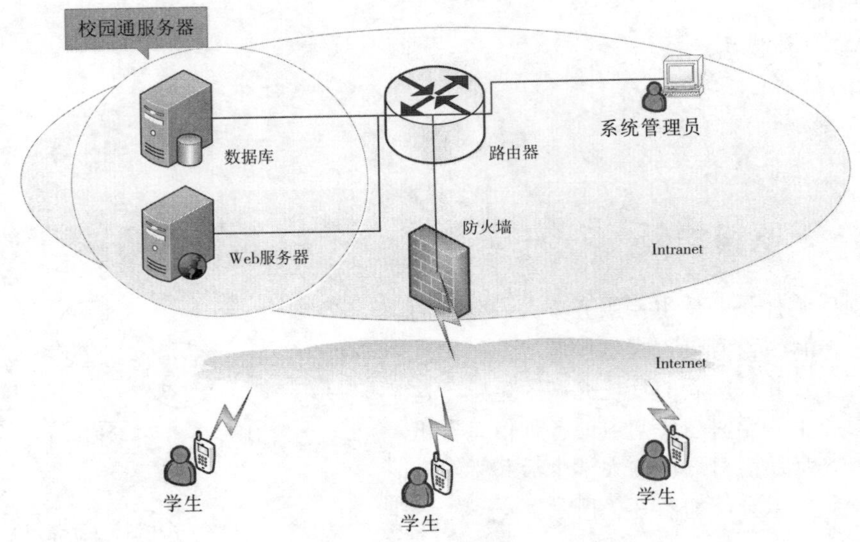

图 8-2 校园通项目物理架构设计图

Android 软件功能架构如图 8-3 所示。

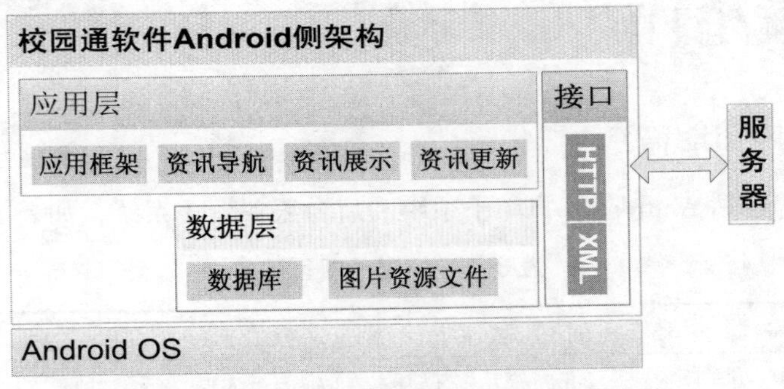

图 8-3 校园通软件 Android 侧架构设计图

Android 侧架构总共分为三层：

（1）应用层负责整个软件的界面展现，通过 DAO 和数据库交互，通过 Android 的 AssertManager 和资源 ID 直接操作图片资源文件，并通过接口层和服务器通信，从而更新数据到数据层。

（2）数据层负责本地业务数据存储。

（3）接口层通过 HTTP+XML 和服务器进行交互。

资讯更新业务流程如下：

（1）通过接口层去服务器获取比目前资讯更新（创建时间更晚）的数据，文本数据放入数据库，图片数据放在本地文件系统中。

（2）应用层通过 DAO 访问数据库中的文本数据，通过 Android 的 AssertManager 和资源 ID 访问图片资源。

1. 项目概要设计

基于总体设计，包设计如表 8-2 所示。

表 8-2 校园通项目包设计说明

包　　名	说　　明
edu.sziit.common	该包及其下的子包中的类和接口，是从本项目抽出来的公共模块，具有很强的通用性，不仅可用于本项目，其他项目中也可以使用
edu.sziit.common.sao	包含服务器对象访问框架，实现了从服务器访问数据并封装为对象的框架，应用通过实现其定义的接口，可以实现具体对象的封装逻辑
edu.sziit.common.sqlite	包含和 SQLite 数据库交互的通用类及接口
edu.sziit.common.widget	包含通用的控件
edu.sziit.infa	包括只能用于校园通项目的类
edu.sziit.infa.adapter	包含项目中所有控件的适配器类的实现
edu.sziit.infa.dao	包含主要的 DAO 实现类，通过该包下的类，可以以对象的方式对数据进行读/写
edu.sziit.infa.pojo	项目中使用到的 Java 对象
edu.sziit.infa.sao	包含服务器数据组装成 Java 对象的实现逻辑，实现了 edu.sziit.common.sao 中定义的接口
edu.sziit.infa.task	包含本项目中使用到的异步任务类，继承 Android 的 AsyncTask 类
edu.sziit.infa.ui	包含所有页面的实现类 Activity
edu.sziit.infa.util	包含项目中使用到的一些公用方法

大体划分原则，首先按照子系统划分顶层包，如本项目顶层包有两个：edu.sziit.common 和 edu.sziit.infa，分别是公共类库和 Android 侧软件，然后在其下，按照类的功能类型划分子包，例如，edu.sziit.infa 顶层包下，划分 edu.sziit.infa.adapter 用于放置所有控件的适配器类的实现类，划分 edu.sziit.infa.ui 用于放置所有页面的实现类 Activity。

在设计过程中，对于比较通用的功能，一定要抽出来做成公共类库，以方便后续项目的开发，避免重复“造轮子”。这也是整个项目团队最珍贵的积累。比如，本项目中就把 SQLite 数据库交互、服务器对象访问以及一些公共控件放到这个包下。类设计如表 8-3 所示。

表 8-3 校园通项目类设计说明

类　　名	说　　明
edu.sziit.common.sqlite.SqliteDaoTool	执行 SQL 语句并返回结果，当结果以对象形式返回时，需要实现 ISqliteRowMapper
edu.sziit.common.sqlite.ISqliteRowMapper	将数据库查询结果封装为对象
edu.sziit.common.sqlite.SqliteDBHelper	所有 SQLite 数据库操作的入口类，包括获得 SqliteDaoTool、上传数据库文件等

续表

类　名	说　明
edu.sziit.common.sao.SaoTool	实现服务器数据访问的框架，用户需实现 IXmlMapper 来组装服务器返回的数据为对象
edu.sziit.common.sao.IXmlMapper	定义了组装服务器返回的数据为对象的接口
edu.sziit.common.widget.MyListView	自定义列表视图，除基本的列表展示功能外，还实现了列表下拉刷新
edu.sziit.infa.ui.WelcomeActivity	欢迎页面的 Activity 类
res/layout/activity_welcome.xml	欢迎页面的布局文件
edu.sziit.infa.ui.GuideGalleryActivity	向导页面的 Activity 类
res/layout/activity_guide_gallery.xml	向导页面的布局文件
edu.sziit.infa.ui.MainActivity	主页面的 Activity 类，包括资讯分类导航和列表展示
res/layout/activity_main.xml	主页面的布局文件
edu.sziit.infa.ui.ZxDetailActivity	单个资讯详细信息展示页面的 Activity 类，包括资讯标题、多图 Gallery 展示以及资讯文本内容展示；点击主页面列表中某项时，跳转到该页面
res/layout/activity_zxdetail.xml	资讯详细信息展示页面的布局文件
edu.sziit.infa.ui.FullScreenImageActivity	资讯图片信息全屏展示页面，点击资讯详细信息展示页面的图片，跳转到该页面
edu.sziit.infa.adapter.GuideGalleryAdapter	向导页面 Gallery 控件适配器类
edu.sziit.infa.adapter.DetailGalleryAdapter	资讯详细信息展示页面 Gallery 控件适配器类
edu.sziit.infa.adapter.ListAdapter	主页面列表控件适配器类
edu.sziit.infa.dao.InfaDaoTool	以对象形式对数据库表的 DML 操作
edu.sziit.infa.dao.InfaDao	InfaDaoTool 的子类，加入复杂数据库操作的方法
edu.sziit.infa.pojo.IFW_ZIXUNPINGTAI_	资讯平台对象数据库映射类，其属性和数据库表字段一一对应
edu.sziit.infa.pojo.IFW_ZIXUNPINGTAI	资讯平台对象数据库映射类的子类,用于增加用户自定义的方法和属性
edu.sziit.infa.pojo.IFW_ZIXUNFENLEI_	资讯分类对象数据库映射类，其属性和数据库表字段一一对应
edu.sziit.infa.pojo.IFW_ZIXUNFENLEI	资讯分类对象数据库映射类的子类,用于增加用户自定义的方法和属性
edu.sziit.infa.pojo.IFW_ZIXUN_	资讯对象数据库映射类，其属性和数据库表字段一一对应
edu.sziit.infa.pojo.IFW_ZIXUN	资讯对象数据库映射类的子类，用于增加用户自定义的方法和属性
edu.sziit.infa.pojo.IFW_BANBENJILU_	软件版本对象数据库映射类，其属性和数据库表字段一一对应
edu.sziit.infa.pojo.IFW_BANBENJILU	软件版本对象数据库映射类的子类,用于增加用户自定义的方法和属性
edu.sziit.infa.sao.ZiXunMapper	将服务器获取到的资讯数据组装成 IFW_ZIXUN 对象
edu.sziit.infa.task.FenleiFromDB	从数据库获取资讯分类数据的异步任务类
edu.sziit.infa.task.ZixunFromDB	从数据库获取资讯数据的异步任务类
edu.sziit.infa.task.ZixunFromServer	从服务器获取资讯数据的异步任务类
edu.sziit.infa.util.InfaUtil	整个项目的工具类，该类如果过大，可以按照其功能划分成不同的工具类

2. 数据库设计

数据库设计如表 8-4～表 8-7 所示。

表 8-4　资讯平台表 IFW_ZIXUNPINGTAI 设计

字　段　名	说　　明	类　　型
M_ID	ID	VARCHAR
M_MC	名称	VARCHAR
M_LOGO	LOGO 文件	VARCHAR
M_UI	UI	VARCHAR
M_GJZ	关键字	VARCHAR
M_CJSJ	创建时间	DATETIME
M_GXSJ	更新时间	DATETIME
M_ORG	组织	VARCHAR
M_V1	备用字段 1	VARCHAR
M_V2	备用字段 2	VARCHAR
M_V3	备用字段 3	VARCHAR
M_CHUANGJIANREN_ID	创建人	VARCHAR
M_XIUGAIREN_ID	修改人	VARCHAR

表 8-5　资讯分类表 IFW_ZIXUNFENLEI 设计

M_NAME	说　　明	类　　型
M_ID	ID	VARCHAR
M_MC	名称	VARCHAR
M_LOGO	LOGO	VARCHAR
M_URL	URL	VARCHAR
M_GJZ	关键字	VARCHAR
M_SXSL	刷新数量	INTEGER
M_SSSX	实时刷新	VARCHAR
M_SY	索引	INTEGER
M_CJSJ	创建时间	DATETIME
M_GXSJ	更新时间	DATETIME
M_ORG	组织	VARCHAR
M_V1	备用字段 1	VARCHAR
M_V2	备用字段 2	VARCHAR
M_V3	备用字段 3	VARCHAR
M_ZIXUNPINGTAI_ID	外键，指向资讯平台表	VARCHAR
M_CHUANGJIANREN_ID	创建人	VARCHAR
M_XIUGAIREN_ID	修改人	VARCHAR

表 8-6　资讯表 IFW_ZIXUN 设计

M_NAME	说　明	类　型
M_JT	静态	VARCHAR
M_NR	内容	VARCHAR
M_URL	URL	VARCHAR
M_GJZ	关键字	VARCHAR
M_CJSJ	创建时间	DATETIME
M_GXSJ	更新时间	DATETIME
M_ORG	组织	VARCHAR
M_V1	备用字段 1	VARCHAR
M_V2	备用字段 2	VARCHAR
M_V3	备用字段 3	VARCHAR
M_ZIXUNFENLEI_ID	外键，指向资讯分类表	VARCHAR
M_CHUANGJIANREN_ID	创建人	VARCHAR
M_XIUGAIREN_ID	修改人	VARCHAR
M_ID	ID	VARCHAR
M_XT	小图	VARCHAR
M_MC	名称	VARCHAR
M_T1	图 1	VARCHAR
M_T2	图 2	VARCHAR
M_T3	图 3	VARCHAR

表 8-7　软件版本表 IFW_BANBENJILU 设计

M_NAME	说　明	类　型
M_ID	ID	VARCHAR
M_BBH	版本号	VARCHAR
M_FJ	附件	VARCHAR
M_BZ	备注	VARCHAR
M_CJSJ	创建时间	DATETIME
M_GXSJ	更新时间	DATETIME
M_ORG	组织	VARCHAR
M_V1	备用字段 1	VARCHAR
M_V2	备用字段 2	VARCHAR
M_V3	备用字段 3	VARCHAR
M_ZIXUNPINGTAI_ID	资讯平台	VARCHAR
M_CHUANGJIANREN_ID	创建人	VARCHAR
M_XIUGAIREN_ID	修改人	VARCHAR

3. **服务器和 Android 侧数据接口设计**

获取最新资讯 URL：

http://localhost:8080/IFW/MOS/engine/loadGridDataXml.mo?mosCrud=IFW_ZIXUN&mosUserNoLogin=ifwuser&SEARCH_IFW_ZIXUN_M_ZIXUNFENLEI_ID=ZIXUNFENLEI-20130716151151-ifwuser&SEARCH_IFW_ZIXUN_M_CJSJ=2013-07-23 23:59:59

参数说明如下：

localhost 为服务器 IP 地址；

mosCrud 参数值参照下面的赋值即可：mosCrud=IFW_ZIXUN；

资讯平台的创建用户（ifwuser）：mosUserNoLogin=ifwuser；

资讯分类 ID（ZIXUNFENLEI-20130716151151-ifwuser）：SEARCH_IFW_ZIXUN_M_ZIXUNFENLEI_ID=ZIXUNFENLEI-20130716151151-ifwuser；

创建时间：SEARCH_IFW_ZIXUN_M_CJSJ=2013-07-23 23:59:59；

页数（0 表示第一页）：start=0；

单页记录限制（每页最多显示 10 条记录）：limit=10；

服务器返回的资讯数据如下：

源代码：/INFA-6/src/edu/sziit/common/sqlite/SqliteDBHelper.java。

```
<?xml version="1.0" encoding="utf-8" ?>
<crud rows="1">
<row>
   <LIST_IFW_ZIXUN_M_ID>ZIXUN-20130716160533-ifwuser</LIST_IFW_ZIXUN_M_ID>
   <LIST_IFW_ZIXUN_M_XT>upload/286e50b1405694274827cc8907ae46375.jpg
</LIST_IFW_ZIXUN_M_XT>
   <LIST_IFW_ZIXUN_M_MC>新生初到谨防被骗</LIST_IFW_ZIXUN_M_MC>
   <LIST_IFW_ZIXUN_M_JT></LIST_IFW_ZIXUN_M_JT>
   <LIST_IFW_ZIXUN_M_NR>        因为是第一次，所以许多同学并不清楚该准备哪些东西，无论是精神上还是物质上。为此，这里为你提供一些\u201c 入学须知\u201d，希望能给大家有所参考。证件材料及早办妥：\n1.户口簿及身份证。入学报到必须交验身份证，因此，当你接到录取通知书后，应尽早凭录取通知书并带上户口簿和身份证到户口所在地派出所申报迁出登记，派出所审查复核后，签发户口迁移证明并注销原户口。根据最新政策精神，被高校录取的新生户口可以不迁，仍留在原地，也可以迁到高校所在地。迁与不迁，由大学新生自己决定。身份证应随身携带入校。\n2.办妥党、团关系的转接手续。如果你是团员身份，应由团籍所在学校团委在团员证上填写转至某高校团委并加盖公章即可。如果是党员身份的同学，则需要所在学校党组织开具转出证明。\n3.其他。新生应当认真阅读录取通知书中的各项要求，并认真办妥。有的学校要求上缴照片，虽然到校后仍可以去拍，但人生地不熟太费时，会白白浪费时间。到校报到谨防骗子，学校入学通知书上都会注明到校时间，最好是在规定的时间里到校，并尽量早到，准备起来会充分一些。趁着假期充分了解学校所在地的文化、风俗习惯、饮食特点、语言、气候等等很有必要，如果有条件的话，最好在父母的陪同下到校体验一次。回家之后就可以分析可能遇到的不适应。新生凭入学通知书可以购买半价火车票，仅限硬座。行李尽量托运，但一些到校后急用的东西，特别是现金、证件材料、托运行李领取单等不要托运。大学都会安排专人在车站码头醒目的地方，设立一个新生接待站，负责接新生入学。学生到站后，应当在出站口找到学校的接待人员，由他们安排车辆送自己到学校。要防止一些假冒高校接送人员的骗子，不要在半路上将钱交给别人。</LIST_IFW_ZIXUN_M_NR>
   <LIST_IFW_ZIXUN_M_URL></LIST_IFW_ZIXUN_M_URL>
   <LIST_IFW_ZIXUN_M_GJZ>新生须知</LIST_IFW_ZIXUN_M_GJZ>
```

```
        <LIST_IFW_ZIXUN_M_CJSJ>2013-07-24 01:15:00</LIST_IFW_ZIXUN_M_CJSJ>
        <LIST_IFW_ZIXUN_M_GXSJ>2013-08-08 09:22:03</LIST_IFW_ZIXUN_M_GXSJ>
        <LIST_IFW_ZIXUN_M_ORG>ORG-DEFAULT</LIST_IFW_ZIXUN_M_ORG>
        <LIST_IFW_ZIXUN_M_V1></LIST_IFW_ZIXUN_M_V1>
        <LIST_IFW_ZIXUN_M_V2></LIST_IFW_ZIXUN_M_V2>
        <LIST_IFW_ZIXUN_M_V3></LIST_IFW_ZIXUN_M_V3>
        <LIST_IFW_ZIXUN_M_T1>upload/2576abb9dc6664a4c9b6acafea647ea7c.jpg
</LIST_IFW_ZIXUN_M_T1>
        <LIST_IFW_ZIXUN_M_T2></LIST_IFW_ZIXUN_M_T2>
        <LIST_IFW_ZIXUN_M_T3></LIST_IFW_ZIXUN_M_T3>
        <LIST_IFW_ZIXUN_M_ZIXUNFENLEI_ID>ZIXUNFENLEI-20130716151151-ifwuser
</LIST_IFW_ZIXUN_M_ZIXUNFENLEI_ID>
        <LIST_IFW_ZIXUN_M_ZIXUNFENLEI_ID0>入学指南</LIST_IFW_ZIXUN_M_ZIXUNFENLEI_ID0>
        <LIST_IFW_ZIXUN_M_CHUANGJIANREN_ID>ifwuser</LIST_IFW_ZIXUN_M_CHUANGJ
IANREN_ID>
        <LIST_IFW_ZIXUN_M_CHUANGJIANREN_ID0>ifwuser</LIST_IFW_ZIXUN_M_CHUANG
JIANREN_ID0>
        <LIST_IFW_ZIXUN_M_XIUGAIREN_ID>ifwuser</LIST_IFW_ZIXUN_M_XIUGAIREN_ID>
        <LIST_IFW_ZIXUN_M_XIUGAIREN_ID0>ifwuser</LIST_IFW_ZIXUN_M_XIUGAIREN_ID0>
    </row>
    </crud>
```

crud 结点 rows 属性表示返回记录的个数；

row 结点各子结点对应的数据说明如下：

```
LIST_IFW_ZIXUN_M_ID    ID
LIST_IFW_ZIXUN_M_XT    小图
LIST_IFW_ZIXUN_M_MC    名称
LIST_IFW_ZIXUN_M_JT    静态
LIST_IFW_ZIXUN_M_NR    内容
LIST_IFW_ZIXUN_M_URL   URL
LIST_IFW_ZIXUN_M_GJZ   关键字
LIST_IFW_ZIXUN_M_CJSJ  创建时间
LIST_IFW_ZIXUN_M_GXSJ  更新时间
LIST_IFW_ZIXUN_M_ORG   ORG
LIST_IFW_ZIXUN_M_V1    V1
LIST_IFW_ZIXUN_M_V2    V2
LIST_IFW_ZIXUN_M_V3    V3
LIST_IFW_ZIXUN_M_ZIXUNFENLEI_ID          资讯分类 ID
LIST_IFW_ZIXUN_M_ZIXUNFENLEI_ID0         资讯分类
LIST_IFW_ZIXUN_M_CHUANGJIANREN_ID        创建人 ID
LIST_IFW_ZIXUN_M_CHUANGJIANREN_ID0       创建人
LIST_IFW_ZIXUN_M_XIUGAIREN_ID            修改人 ID
LIST_IFW_ZIXUN_M_XIUGAIREN_ID0           修改人
LIST_IFW_ZIXUN_M_T1    图 1
LIST_IFW_ZIXUN_M_T2    图 2
LIST_IFW_ZIXUN_M_T3    图 3
```

8.3 必备的技术和知识点

8.3.1 DAO 设计模式

DAO（Data Access Object，数据访问对象）是一个数据访问接口，数据访问：顾名思义就是与数据库打交道，夹在业务逻辑与数据库资源中间。

在核心 J2EE 模式中是这样介绍 DAO 模式的：为了建立一个健壮的 J2EE 应用，应该将所有对数据源的访问操作抽象封装在一个公共 API 中。用程序设计语言来说，就是建立一个接口，接口中定义了此应用程序中将会用到的所有事务方法。在这个应用程序中，当需要和数据源进行交互时则使用这个接口，并且编写一个单独的类来实现这个接口在逻辑上对应这个特定的数据存储。

DAO 模式是标准的 J2EE 设计模式之一，开发人员使用这个模式把底层的数据访问操作和上层的商务逻辑分开。一个典型的 DAO 实现有下列几个组件：

（1）一个 DAO 工厂类。

（2）一个 DAO 接口。

（3）一个实现 DAO 接口的具体类。

（4）数据传递对象（有时称为值对象）。

本项目中，对此设计模式做了简化，没有定义 DAO 接口，InfaDao 是实现 DAO 接口职责的具体类。

InfaUtil 类的 getReadableDao()和 getWritebleDao()方法，承担了 DAO 工厂类的职责。

包 edu.sziit.infa.pojo 下的所有类就是数据传递对象。

因此，本项目采用了 DAO 设计思想，而做了简化的实现。

8.3.2 XML 解析

XML 在各种开发中都广泛应用，Android 也不例外。作为承载数据的一个重要角色，如何读/写 XML 成为 Android 开发中一项重要的技能。接下来向大家介绍在 Android 平台下几种常见的 XML 解析和创建方法。

在 Android 中，常见的 XML 解析器分别为 SAX 解析器、DOM 解析器和 PULL 解析器。

1. SAX 解析器

SAX（Simple API for XML）解析器是一种基于事件的解析器，它的核心是事件处理模式，主要是围绕着事件源以及事件处理器来工作的。当事件源产生事件后，调用事件处理器相应的处理方法，一个事件就可以得到处理。在事件源调用事件处理器中特定方法时，还要传递给事件处理器相应事件的状态信息，这样事件处理器才能根据提供的事件信息决定自己的行为。

SAX 解析器的优点是解析速度快，占用内存少。非常适合在 Android 移动设备中使用。

2. DOM 解析器

DOM 是基于树形结构的结点或信息片段的集合，允许开发人员使用 DOM API 遍历 XML 树、检索所需数据。分析该结构通常需要加载整个文档和构造树形结构，然后才可以检索和更新结点信息。

由于 DOM 在内存中以树形结构存放，因此检索和更新效率会更高。但是对于特别大的文档，解析和加载整个文档将会很耗资源。

3. PULL 解析器

PULL 解析器的运行方式和 SAX 类似，都是基于事件的模式。不同的是，在 PULL 解析过程中，用户需要自己获取产生的事件然后做相应的操作，而不像 SAX 那样由处理器触发一种事件的方法，执行用户的代码。PULL 解析器小巧轻便，解析速度快，简单易用，非常适合在 Android 移动设备中使用，Android 系统内部在解析各种 XML 时也是用 PULL 解析器。

本项目中采用 PULL 解析器，具体步骤如下：

（1）获得解析器工厂 XmlPullParserFactory。

（2）获得解析器 XmlPullParser，并设置 XML 输入流。

（3）在 while 循环中不断调用解析器的 next()方法，根据不同的事件类型进行解析。

具体代码参见 INFA-8\INFA\src\edu\sziit\common\sao\SaoTool.java 和 INFA-8\INFA\src\edu\sziit\infa\sao\ZiXunMapper.java（粗体部分）：

源代码：INFA-8\INFA\src\edu\sziit\common\sao\SaoTool.java。

```
/**
   * 从服务器获取对象数据
   * @param xmlPath http 请求的全路径 url
   * @param mapper IXmlMapper 接口的实现，组装 xmlPath 对应请求的 XML 数据为对象
   * @return
   */
  public static <T> List<T> getList(String xmlPath, IXmlMapper<T> mapper) {
     InputStream inputStream=null;
     List<T> zixuns=new ArrayList<T>();
     HttpURLConnection urlConn=null;
     URL url=null;
     try {
        //建立连接获取数据
        url=new URL(xmlPath);
        urlConn=(HttpURLConnection) url.openConnection();
        urlConn.setDoInput(true);
        urlConn.setDoOutput(true);
        urlConn.setUseCaches(false);
        urlConn.setRequestMethod("POST");
        urlConn.setRequestProperty("Content-Type", "text/xml;charset=utf-8");
        inputStream=urlConn.getInputStream();
        // 获取 XmlPullParser，解析数据
        XmlPullParserFactory factory=XmlPullParserFactory.newInstance();
        XmlPullParser xmlParser=factory.newPullParser();
        xmlParser.setInput(inputStream, "UTF-8");
        return mapper.parseList(xmlParser);
     } catch (Exception e) {
        Log.e(TAG, "get data from server failed...", e);
        e.printStackTrace();
     } finally {
        try {
```

```
            inputStream.close();
        } catch (Exception e) {
        }
        try {
            urlConn.disconnect();
        } catch (Exception e) {
        }
    }
    return zixuns;
}
```

源代码：INFA-8\INFA\src\edu\ sziit\infa\sao\ZiXunMapper.java。

```
//从服务器获取资讯数据，并封装为 List<IFW_ZIXUN>
public List<IFW_ZIXUN> parseList(XmlPullParser xmlParser) {
    List<IFW_ZIXUN> zixuns=new ArrayList<IFW_ZIXUN>();
    IFW_ZIXUN zixun=null;
    try {
        int evtType=xmlParser.getEventType();
        while (evtType != XmlPullParser.END_DOCUMENT) {
            switch (evtType) {
            case XmlPullParser.START_DOCUMENT:
                break;
            case XmlPullParser.START_TAG:
                if (xmlParser.getName().equals("row")) {
                    zixun = new IFW_ZIXUN();
                } else if ("LIST_IFW_ZIXUN_M_ID".equals(xmlParser.getName())) {
                    zixun.setM_ID(xmlParser.nextText());
                } else if ("LIST_IFW_ZIXUN_M_XT".equals(xmlParser.getName())) {
                    zixun.setM_XT(xmlParser.nextText());
                } else if ("LIST_IFW_ZIXUN_M_MC".equals(xmlParser.getName())) {
                    zixun.setM_MC(xmlParser.nextText());
                } else if ("LIST_IFW_ZIXUN_M_NR".equals(xmlParser.getName())) {
                    zixun.setM_NR(xmlParser.nextText());
                } else if ("LIST_IFW_ZIXUN_M_GJZ".equals(xmlParser.getName())) {
                    zixun.setM_GJZ(xmlParser.nextText());
                } else if ("LIST_IFW_ZIXUN_M_CJSJ".equals(xmlParser.getName())) {
                    zixun.setM_CJSJ(xmlParser.nextText());
                } else if ("LIST_IFW_ZIXUN_M_GXSJ".equals(xmlParser.getName())) {
                    zixun.setM_GXSJ(xmlParser.nextText());
                } else if ("LIST_IFW_ZIXUN_M_ORG".equals(xmlParser.getName())) {
                    zixun.setM_ORG(xmlParser.nextText());
                } else if ("LIST_IFW_ZIXUN_M_T1".equals(xmlParser.getName())) {
                    zixun.setM_T1(xmlParser.nextText());
                } else if ("LIST_IFW_ZIXUN_M_T2".equals(xmlParser.getName())) {
                    zixun.setM_T2(xmlParser.nextText());
                } else if ("LIST_IFW_ZIXUN_M_T3".equals(xmlParser.getName())) {
                    zixun.setM_T3(xmlParser.nextText());
```

```
                    } else if ("LIST_IFW_ZIXUN_M_ZIXUNFENLEI_ID".equals(xmlParser
.getName())) {
                        zixun.setM_ZIXUNFENLEI_ID(xmlParser.nextText());
                    } else if ("LIST_IFW_ZIXUN_M_XIUGAIREN_ID".equals(xmlParser
.getName())) {
                        zixun.setM_XIUGAIREN_ID(xmlParser.nextText());
                    }
                    break;
                case XmlPullParser.END_TAG:
                    if (xmlParser.getName().equals("row")) {
                        zixuns.add(zixun);
                        zixun=null;
                    }
                    break;
                default:
                    break;
                }
                evtType = xmlParser.next();
            }
        } catch (Exception e) {
            Log.e(TAG, "Parsing IFW_ZIXUN error", e);
        }
        return zixuns;
    }
```

8.3.3 服务器数据访问

本项目中服务器数据访问采用 HTTP，并且采用 POST 方法向服务器请求数据，具体步骤如下：

（1）构造 URL 获取 HTTP 连接。

（2）设置连接参数，将请求方法设置为 POST。

（3）获取输入流读取服务器数据。

具体代码参见 INFA-8\INFA\src\edu\sziit\common\sao\SaoTool.java（粗体部分）:

源代码：INFA-8\INFA\src\edu\sziit\common\sao\SaoTool.java。

```
/**
   * 从服务器获取对象数据
   * @param xmlPath http请求的全路径url
   * @param mapper IXmlMapper接口的实现，组装xmlPath对应请求的XML数据为对象
   * @return
   */
  public static <T> List<T> getList(String xmlPath, IXmlMapper<T> mapper) {
      InputStream inputStream=null;
      List<T> zixuns=new ArrayList<T>();
      HttpURLConnection urlConn=null;
      URL url=null;
      try {
          //建立连接获取数据
```

```java
            url=new URL(xmlPath);
            urlConn=(HttpURLConnection) url.openConnection();
            urlConn.setDoInput(true);
            urlConn.setDoOutput(true);
            urlConn.setUseCaches(false);
            urlConn.setRequestMethod("POST");
            urlConn.setRequestProperty("Content-Type", "text/xml;charset=utf-8");
            inputStream=urlConn.getInputStream();
            // 获取 XmlPullParser，解析数据
            XmlPullParserFactory factory = XmlPullParserFactory.newInstance();
            XmlPullParser xmlParser = factory.newPullParser();
            xmlParser.setInput(inputStream, "UTF-8");
            return mapper.parseList(xmlParser);
        } catch (Exception e) {
            Log.e(TAG, "get data from server failed...", e);
            e.printStackTrace();
        } finally {
            try {
                inputStream.close();
            } catch (Exception e) {
            }
            try {
                urlConn.disconnect();
            } catch (Exception e) {
            }
        }
        return zixuns;
    }
```

HTTP 请求方法还可以使用 GET，它和 POST 方法的区别如下：

POST 请求可以向服务器传送数据，而且数据放在 HTML HEADER 内一起传送到服务端 URL 地址，数据对用户不可见。而 GET 是把参数数据队列加到提交的 URL 中，值和表单内各个字段一一对应，例如（http://www.baidu.com/s?w=%C4&inputT=2710）。

GET 传送的数据量较小，不能大于 2 KB。POST 传送的数据量较大，一般被默认为不受限制。但理论上，IIS4 中最大量为 80 KB，IIS5 中为 100 KB。

GET 安全性非常低，POST 安全性较高。

8.3.4 异步任务 AsyncTask

Android 的 AsyncTask 比 Handler 更轻量级一些，适用于简单的异步处理。

Handler 和 AsyncTask 都是为了不阻塞主线程（UI 线程），且 UI 的更新只能在主线程中完成，因此异步处理是不可避免的。

Android 为了降低这个开发难度，提供了 AsyncTask。AsyncTask 就是一个封装过的后台任务类，顾名思义就是异步任务。

AsyncTask 直接继承于 Object 类，位置为 android.os.AsyncTask。要使用 AsyncTask 工作须提供三个泛型参数，并重写最少以下这两个方法：

doInBackground：后台执行，比较耗时的操作都可以放在这里。注意这里不能直接操作UI。此方法在后台线程执行，完成任务的主要工作，通常需要较长的时间。在执行过程中可以调用 publishProgress 更新任务的进度。

onPostExecute：相当于 Handler 处理 UI 的方式，在这里可以使用在 doInBackground 中得到的结果处理操作 UI。此方法在主线程执行，任务执行的结果作为此方法的参数返回。

有必要的话用户还得重写以下三个方法，但不是必需的：

onProgressUpdate：可以使用进度条增加用户体验度。此方法在主线程执行，用于显示任务执行的进度。

onPreExecute：这里是最终用户调用 Excute 时的接口，当任务执行之前开始调用此方法，可以在这里显示进度对话框。

onCancelled：用户调用取消时，要进行的操作。

使用 AsyncTask 类，以下是几条必须遵守的准则：

Task 的实例必须在 UI thread 中创建。

execute 方法必须在 UI thread 中调用。

不要手动地调用 onPreExecute()、onPostExecute(Result)、onProgressUpdate(Progress...)，doInBackground(Params...)这几个方法。

该 task 只能被执行一次，否则多次调用时将会出现异常。

本项目中包 edu.sziit.infa.task 下的类都是 AsyncTask 的实现类，重写了 doInBackground 和 onPostExecute()方法，具体分析如表 8-8 所示。

表 8-8　校园通项目任务类详细说明

任　务　类	doInBackground 说明	onPostExecute 说明
edu.sziit.infa.task.FenleiFromDB	从本地数据库获取资讯分类数据	回调主页面 initListNavigator()方法，设置列表导航栏标题
edu.sziit.infa.task.ZixunFromDB	从本地数据库获取资讯数据	回调主页面 refreshZixunList()方法，刷新列表 UI
edu.sziit.infa.task.ZixunFromServer	从服务器获取更新的资讯数据，文本数据入本地数据库，图片数据存储在 SD 卡	回调主页面 afterZixunRefresh()方法，刷新列表 UI

8.4　项目实施

8.4.1　导入项目所需要的资源

导入项目所需要的资源包括数据库文件和图片资源文件。本项目基于 Android 框架项目，复制 INFA-5/INFA 项目到本地，导入 Eclipse 工作空间，并重命名项目为 INFA-6。

将 INFA-6/data 目录下文件夹复制到项目中，包括 isql 数据库文件 res/raw/infa.db 和 assets/upload/*.png 下所有图片资源文件，如图 8-4 所示。

数据库文件中的数据表按照上节数据库设计建立，并初始化了部分数据，作为测试之用。

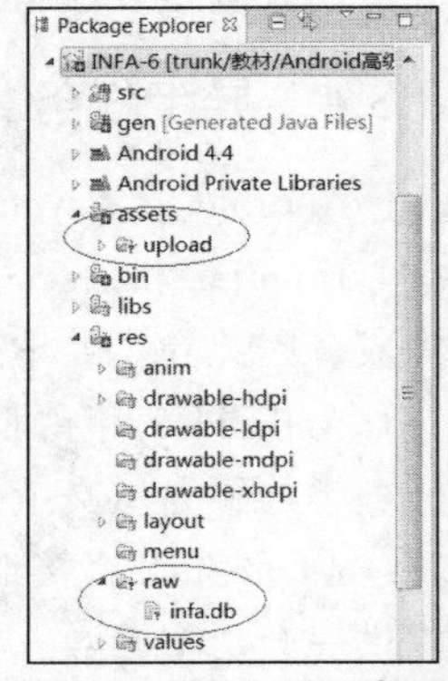

图 8-4　导入项目资源

8.4.2 建立数据库相关的包

创建包 edu.sziit.common.sqlite 实现和 Sqlite 数据库交互的通用类及接口。同时创建如下类：

edu.sziit.common.sqlite.SqliteDaoTool：执行 SQL 语句并返回结果，当结果以对象形式返回时，需要实现 ISqliteRowMapper。

edu.sziit.common.sqlite.ISqliteRowMapper：将数据库查询结果封装为对象。

edu.sziit.common.sqlite.SqliteDBHelper：所有 Sqlite 数据库操作的入口类，包括获得 SqliteDaoTool、上传数据库文件等，如图 8-5 所示。源代码如下：

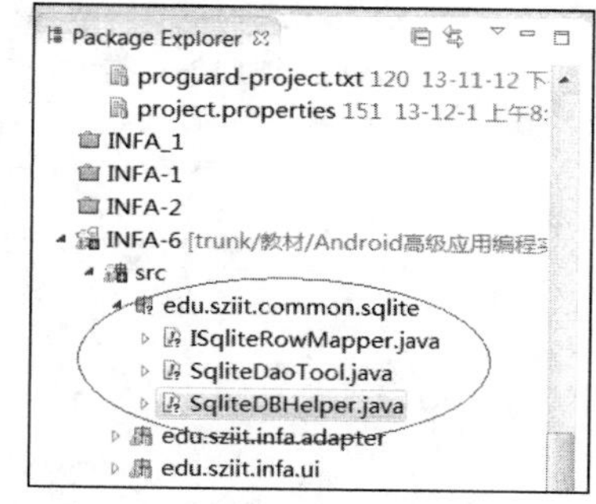

图 8-5　创建 sqlite 包及类

源代码：/INFA-6/src/edu/sziit/common/sqlite/SqliteDBHelper.java。

```
/*** 所有 Sqlite 数据库操作的入口类，包括获得 SqliteDaoTool、上传数据库文件等*/
public class SqliteDBHelper {
   private SQLiteOpenHelper openHelper=null;
   public SqliteDBHelper(Context context, String dbName) {    //构造函数
      openHelper=new SQLiteOpenHelper(context, dbName, null, 1) {
         public void onCreate(SQLiteDatabase db) {
         }
         public void onUpgrade(SQLiteDatabase db, int oldVersion,int newVersion){
         }
      };
   }
   public SqliteDBHelper(SQLiteOpenHelper openHelper) {       //构造函数
      if(openHelper==null) {
         throw new RuntimeException("SQLiteOpenHelper is null.");
      }
      this.openHelper=openHelper;
   }
   private SQLiteDatabase getReadableDatabase() {       //获取只读数据库
      if(openHelper==null) {
         throw new RuntimeException("No SQLiteOpenHelper implemented.");
      }
      return openHelper.getReadableDatabase();
   }
   public SQLiteDatabase getWritableDatabase() {        //获取可写数据库
      if(openHelper==null) {
         throw new RuntimeException("No SQLiteOpenHelper implemented.");
      }
      return openHelper.getWritableDatabase();
   }
   public SqliteDaoTool getReadableDaoTool() {    //获取只读 SqliteDaoTool
      return new SqliteDaoTool(getReadableDatabase());
   }
   public SqliteDaoTool getWritableDaoTool() {    //获取可写 SqliteDaoTool
      return new SqliteDaoTool(getWritableDatabase());
```

```
    }
    public static void updateDBFile(Context context,String dataBasePath,int dbRes){
        //上传数据库文件到手机
        File dbfile=new File(dataBasePath);
        File dir=dbfile.getParentFile();
        if(!dir.exists()){
            dir.mkdirs();
        }
        InputStream is=null;
        FileOutputStream fos=null;
        try {
            if (!dbfile.exists()) {
                dbfile.createNewFile();
                is=context.getApplicationContext().getResources().openRawResource
(dbRes);
                fos=new FileOutputStream(dbfile);
                byte[] buff=new byte[is.available()];
                is.read(buff);
                fos.write(buff);
            }
        } catch (Exception e) {
            throw new RuntimeException(e.getMessage());
        } finally {
            try {
                is.close();
            } catch (Exception e) {
            }
            try {
                fos.close();
            } catch (Exception e) {
            }
        }
    }
}
```

源代码：/INFA-6/src/edu/sziit/common/sqlite/SqliteDaoTool.java。

```
package edu.sziit.common.sqlite;
public class SqliteDaoTool {
    //执行SQL语句并返回结果，当结果以对象形式返回时，需要实现ISqliteRowMapper
    private SQLiteDatabase db=null;
    SqliteDaoTool(SQLiteDatabase db) {
        this.db=db;
    }
    public String queryForString(String sql) {
        Cursor c=null;
        try {
            c=db.rawQuery(sql, new String[] {});
            if(c.moveToFirst()) {
                return c.getString(0);
            }
        } finally {
```

```
        try {
            c.close();
        } catch (Exception e) {
        }
    }
    return null;
}
public <T> List<T> query(String sql, ISqliteRowMapper<T> rm) {
    Cursor c=null;
    List<T> list=new ArrayList<T>();
    try {
        c=db.rawQuery(sql, new String[] {});
        int i=0;
        while(c.moveToNext()) {
            T obj=rm.mapRow(c, i);
            list.add(obj);
            i++;
        }
    } finally {
        try {
            c.close();
        } catch (Exception e) {
        }
    }
    return list;
}
public <T> T queryFirst(String sql, ISqliteRowMapper<T> rm) {
    Cursor c=null;
    try {
        c=db.rawQuery(sql, new String[] {});
        if (c.moveToNext()) {
            return rm.mapRow(c, 0);
        }
    } finally {
        try {
            c.close();
        } catch (Exception e) {
        }
    }
    return null;
}
public void execute(String sql) {
    db.execSQL(sql);
}
public long insert(String table, ContentValues cv) {
    return db.insert(table, null, cv);
}
public long delete(String table, String whereClause) {
    return db.delete(table, whereClause, null);
}
```

```
    public long update(String table, ContentValues cv, String whereClause,
String[] whereArgs) {
        return db.update(table, cv, whereClause, whereArgs);
    }
    public void close() {
        try {
            db.close();
        } catch (Exception e) {
        }
    }
}
```

源代码：/INFA-6/src/edu/sziit/common/sqlite/ISqliteRowMapper.java

```
package edu.sziit.common.sqlite;
public interface ISqliteRowMapper<T> {  //将数据库查询结果封装为对象
    T mapRow (Cursor c, int row);       // 将数据库查询结果封装为类型为 T 的对象
}
```


8.4.3 添加 DAO 类和 Java 对象类

创建 edu.sziit.infa.dao 包，包含主要的 DAO 实现类，通过该包下的类，可以以对象的形式对数据进行读/写，包含的类如下：

edu.sziit.infa.dao.InfaDaoTool：以对象的形式对数据库表进行 DML 操作。

edu.sziit.infa.dao.InfaDao：InfaDaoTool 的子类，加入复杂数据库操作的方法。

具体类请参见 INFA-6/INFA/src/edu/sziit/infa/dao 目录下的.java 文件。

创建 edu.sziit.infa.pojo 包，该包包含项目中使用到的 Java 对象如下：

edu.sziit.infa.pojo.IFW_ZIXUNPINGTAI_：资讯平台对象数据库映射类，其属性和数据库表字段一一对应。

edu.sziit.infa.pojo.IFW_ZIXUNPINGTAI：资讯平台对象数据库映射类的子类，用于增加用户自定义的方法和属性。

edu.sziit.infa.pojo.IFW_ZIXUNFENLEI_：资讯分类对象数据库映射类，其属性和数据库表字段一一对应。

edu.sziit.infa.pojo.IFW_ZIXUNFENLEI：资讯分类对象数据库映射类的子类，用于增加用户自定义的方法和属性。

edu.sziit.infa.pojo.IFW_ZIXUN_：资讯对象数据库映射类，其属性和数据库表字段一一对应。

edu.sziit.infa.pojo.IFW_ZIXUN：资讯对象数据库映射类的子类，用于增加用户自定义的方法和属性。

edu.sziit.infa.pojo.IFW_BANBENJILU_：软件版本对象数据库映射类，其属性和数据库表字段一一对应。

edu.sziit.infa.pojo.IFW_BANBENJILU：软件版本对象数据库映射类的子类，用于增加用户自定义的方法和属性。

具体类请参见 INFA-6/INFA/src/edu/sziit/infa/pojo 目录下的.java 文件。

8.4.4 自定义列表视图，实现下拉刷新功能

图 8-6 列表视图布局图

添加 edu.sziit.common.widget.MyListView 类，实现自定义 ListView 子类，实现列表下拉刷新功能。下拉刷新接口为 MyListView.RefreshListener。实现该接口即可以在下拉刷新时加入自定义代码更新 ListView 视图。具体代码参见 INFA-6\INFA\src\edu\sziit\common\widget\ MyListView.java。

添加该类之前，需要添加布局文件 list_head.xml 到 res/layout/目录下，该布局实现了下拉列表到顶时，显示刷新数据的提示表头（如图 8-6 箭头 1 所指）。

源代码：list_head.xml。

```
<?xml version="1.0" encoding="utf-8"?>
<!-- ListView 的头部 -->
<LinearLayout
xmlns:android="http://schemas.android.com/apk/res/android"
    android:layout_width="fill_parent"
    android:layout_height="wrap_content">
    <!-- 内容 -->
    <RelativeLayout
        android:id="@+id/head_contentLayout"
        android:layout_width="fill_parent"
        android:layout_height="wrap_content"
        android:paddingLeft="30dp" >
        <!-- 箭头图像、进度条 -->
        <FrameLayout
            android:layout_width="wrap_content"
            android:layout_height="wrap_content"
            android:layout_alignParentLeft="true"
            android:layout_centerVertical="true" >
            <!-- 箭头 -->
            <ImageView
                android:id="@+id/list_head_iv_arrow"
                android:layout_width="wrap_content"
                android:layout_height="wrap_content"
                android:layout_gravity="center"
                android:src="@drawable/list_head_arrow" />
            <!-- 进度条 -->
            <ProgressBar
                android:id="@+id/list_head_pb"
                style="?android:attr/progressBarStyleSmall"
                android:layout_width="wrap_content"
                android:layout_height="wrap_content"
                android:layout_gravity="center"
                android:visibility="gone" />
        </FrameLayout>
        <!-- 提示、最近更新 -->
        <LinearLayout
```

```
            android:layout_width="wrap_content"
            android:layout_height="wrap_content"
            android:layout_centerHorizontal="true"
            android:gravity="center_horizontal"
            android:orientation="vertical" >
            <!-- 提示 -->
            <TextView
                android:id="@+id/list_head_tv_tips"
                android:layout_width="wrap_content"
                android:layout_height="wrap_content"
                android:text="@string/pull_to_update"
                android:textColor="@color/dimgrey"
                android:textSize="20sp" />
            <!-- 最近更新 -->
            <TextView
                android:id="@+id/list_head_tv_lastupdate"
                android:layout_width="wrap_content"
                android:layout_height="wrap_content"
                android:text="@string/last_update"
                android:textColor="@color/dimgrey"
                android:textSize="10sp" />
        </LinearLayout>
    </RelativeLayout>
</LinearLayout>
```

该布局引用的图片、文字及颜色资源，请参看 INFA-6\INFA\res 目录下的 drawable-hdpi、values\strings.xml 以及 values\colors.xml。该布局如图 8-6 箭头 1 所指向的列表头。

8.4.5 定义列表适配器类，实现列表展现

之前需要添加 edu.sziit.infa.util.InfaUtil 类，把一些公共方法抽取出来，具体见 INFA-6/src/edu/sziit/infa/util/InfaUtil.java 文件。添加 edu.sziit.infa.adapter.ListAdapter 类，实现列表适配器。其中 getView()方法实现了每一个列表项的展示视图，代码如下：

源代码：edu.sziit.infa.adapter.ListAdapter getView 方法。

```
    /**@see android.widget.Adapter#getView(int, android.view.View, android.view.
ViewGroup)*/
    //实现每一个列表项的展示视图
    public View getView(int position, View convertView, ViewGroup parent) {
        int viewType=getItemViewType(position);
        IFW_ZIXUN zx=zxList.get(position);
        if (convertView==null) {
            if (viewType==0) {
                convertView=View.inflate(context, R.layout.list_item_top,null);
                ViewHolder v=new ViewHolder();
                v.text=InfaUtil.getTextView(convertView,R.id.list_item_top_text);
                v.image=InfaUtil.getImageView(convertView,R.id.list_item_top_image);
                convertView.setTag(v);
            } else {
                convertView=View.inflate(context, R.layout.list_item, null);
                ViewHolder v=new ViewHolder();
                v.text=InfaUtil.getTextView(convertView, R.id.list_item_title);
```

```
            v.image=InfaUtil.getImageView(convertView,R.id.list_item_image);
            convertView.setTag(v);
        }
    }
    ViewHolder v=(ViewHolder) convertView.getTag();
    v.text.setText(zx.getM_MC());
    v.zixun=zx;
    if (viewType==0) {
        InfaUtil.setImageBackground(v.image, zx.getImageList().get(0));
    } else {
        InfaUtil.setImageBackground(v.image, zx.getM_XT());
    }
    return convertView;
}
```

该代码对应两个布局文件 list_item_top.xml 和 list_item.xml，分别对应图 8-6 箭头 2 和 3 所指向的部分，代码如下：

源代码：list_item_top.xml。

```
<?xml version="1.0" encoding="utf-8"?>
<FrameLayout xmlns:android="http://schemas.android.com/apk/res/android"
    android:layout_width="fill_parent"
    android:layout_height="wrap_content" >
    <ImageView
        android:id="@+id/list_item_top_image"
        android:layout_width="fill_parent"
        android:layout_height="220dp" >
    </ImageView>
    <TextView
        android:id="@+id/list_item_top_text"
        android:layout_width="fill_parent"
        android:layout_height="wrap_content"
        android:layout_gravity="bottom"
        android:background="@drawable/list_item_top_text_bg"
        android:gravity="center"
        android:textColor="@color/white"
        android:textSize="20sp" />
</FrameLayout>
```

源代码：list_item.xml。

```
<?xml version="1.0" encoding="UTF-8"?>
<LinearLayout xmlns:android="http://schemas.android.com/apk/res/android"
    android:layout_width="match_parent"
    android:layout_height="wrap_content"
    android:background="@drawable/list_item"
    android:gravity="center_vertical"
    android:minHeight="?android:listPreferredItemHeight"
    android:orientation="horizontal"
    android:paddingBottom="8.0dip"
    android:paddingTop="8.0dip" >
    <LinearLayout
        android:layout_width="80dip"
```

```
        android:layout_height="80dip"
        android:layout_marginLeft="2.0dip" >
        <ImageView
            android:id="@+id/list_item_image"
            android:layout_width="80dip"
            android:layout_height="80dip"
            android:scaleType="centerCrop" >
        </ImageView>
    </LinearLayout>
    <LinearLayout
        android:layout_width="match_parent"
        android:layout_height="wrap_content"
        android:layout_marginLeft="6.0dip"
        android:orientation="vertical"
        android:paddingRight="4.0dip" >
        <LinearLayout
            android:layout_width="fill_parent"
            android:layout_height="wrap_content"
            android:orientation="horizontal" >
            <TextView
                android:id="@+id/list_item_title"
                android:layout_width="wrap_content"
                android:layout_height="wrap_content"
                android:ellipsize="marquee"
                android:singleLine="true"
                android:textAppearance="?android:textAppearanceMedium"
                android:textColor="#ff000000" />
        </LinearLayout>
        <LinearLayout
            android:layout_width="fill_parent"
            android:layout_height="wrap_content"
            android:orientation="horizontal"
            android:visibility="gone" >
            <TextView
                android:id="@+id/list_item_detail"
                android:layout_width="0dp"
                android:layout_height="wrap_content"
                android:layout_weight="1"
                android:ellipsize="end"
                android:includeFontPadding="false"
                android:maxLines="2"
                android:textAppearance="?android:textAppearanceSmall"
                android:textColor="#ff000000"
                android:textSize="14.0sp" />
        </LinearLayout>
    </LinearLayout>
</LinearLayout>
```

布局设计的所有资源请参见 INFA-6/INFA/res 目录；edu.sziit.infa.adapter.ListAdapter 类的完整代码请参见：INFA-6\INFA\src\edu\sziit\infa\adapter\ListAdapter.java

8.4.6 修改主页面的布局和 Activity 类，完成列表展示

之前需要实现两个异步任务，从数据库获取资讯分类和资讯数据，代码详见：

```
INFA-6\INFA\src\edu\sziit\infa\task\FenleiFromDB.java
INFA-6\INFA\src\edu\sziit\infa\task\ZixunFromDB.java
```

修改布局文件 res/layout/activity_main.xml，代码如下：

源代码：edu.sziit.infa.ui.WelcomeActivity。

```
<RelativeLayout xmlns:android="http://schemas.android.com/apk/res/android"
    xmlns:tools="http://schemas.android.com/tools"
    android:layout_width="match_parent"
    android:layout_height="match_parent"
    android:background="@color/white" >
    <HorizontalScrollView
        android:id="@+id/list_navigator"
        android:layout_width="match_parent"
        android:layout_height="38dp"
        android:background="@color/white"
        android:fadingEdge="@null"
        android:scrollbars="none" >
        <RelativeLayout
            android:id="@+id/list_navigator_scroll"
            android:layout_width="wrap_content"
            android:layout_height="match_parent"
            android:background="@color/list_navigator_text_select" >
            <RadioGroup
                android:id="@+id/list_navigator_radioGroup"
                android:layout_width="fill_parent"
                android:layout_height="37dp"
                android:layout_alignParentTop="true"
                android:background="@color/white"
                android:orientation="horizontal" >
            </RadioGroup>
            <ImageView
                android:id="@+id/list_navigator_bottom_line"
                android:layout_width="50dp"
                android:layout_height="4dp"
                android:layout_alignParentBottom="true"
                android:layout_marginLeft="0dp"
                android:background="@color/list_navigator_text_select" />
        </RelativeLayout>
    </HorizontalScrollView>
    <RelativeLayout
        android:id="@+id/list_margin"
        android:layout_width="match_parent"
        android:layout_height="5dp"
        android:layout_below="@+id/list_navigator"
        android:background="@color/white" >
    </RelativeLayout>
    <edu.sziit.common.widget.MyListView
        android:id="@+id/list"
        android:layout_width="match_parent"
        android:layout_height="match_parent"
        android:layout_alignParentLeft="true"
```

```
        android:layout_below="@+id/list_margin"
        android:addStatesFromChildren="true"
        android:cacheColorHint="@color/white"
        android:divider="@color/white" />
</RelativeLayout>
```

该布局在顶部用 RadioGroup 控件做了一个列表导航栏，下部是自定义的列表 MyListView，布局设计的所有资源请参见 INFA-6/INFA/res 目录。修改 src/edu/sziit/infa/ui/MainActivity.java，具体代码参见 INFA-6\INFA\src\edu\sziit\infa\ui\MainActivity.java。大致的程序流程如下：

（1）在 Activity 的 onCreate()方法中，启动 FenleiFromDB 任务从数据库获取资讯分类数据，并回调 MainActivity 的 initListNavigator()方法，初始化导航栏中的所有标题。

（2）然后触发导航栏第一项被单击的事件：

```
listNavigatorRadioGroup.findViewById(0).performClick();
```

（3）在改变导航栏标题的响应方法 onNavigatorChange()中，启动 ZixunFromDB 任务，从数据库获取资讯数据，并回调 MainActivity 的 refreshZixunList()方法，刷新列表数据。

8.4.7 上传数据库文件

在 edu.sziit.infa.ui.WelcomeActivity 初始化方法 onCreate()中增加如下代码（粗体），在页面启动时上传数据库文件到 Android 文件系统中（如果已经上传，则什么也不做）：

源代码：edu.sziit.infa.ui.WelcomeActivity。

```
  protected void onCreate(Bundle savedInstanceState) {   //页面初始化
     super.onCreate(savedInstanceState);
     requestWindowFeature(Window.FEATURE_NO_TITLE);      //不显示标题
     setContentView(R.layout.activity_welcome);          //设置布局
     InfaUtil.uploadDBFile(this);              // 拷贝数据库文件到指定路径
     // 设置图像的动画效果
     Animation animation=AnimationUtils.loadAnimation(this,R.anim. welcome_shade);
     ImageView imageView=(ImageView) findViewById(R.id.welcome);
     imageView.startAnimation(animation);
     // 获取系统参数 isFirst，判断是否第一次启动
     SharedPreferences preferences=getSharedPreferences(this. getPackageName(),
MODE_PRIVATE);
     boolean isFirst=preferences.getBoolean("isFirst", true);
     if(isFirst){
        createShortCut();              // 创建快捷方式
        SharedPreferences.Editor editor = preferences.edit();
                                       //设置参数 isFirst=false
        editor.putBoolean("isFirst", false);
        editor.commit();
        msgHandler.sendEmptyMessageDelayed(GO_GUIDE, flashInterval);
     } else {
        msgHandler.sendEmptyMessageDelayed(GO_HOME, flashInterval);
     }
  }
```

InfaUtil.uploadDBFile 代码如下，将打包后的 apk 中的 raw/infa.db 文件复制到 Android 系统文件夹下面："/data/data/" + context.getPackageName() + "/databases/infa.db"。

注意：infa.db 文件必须放在/data/data/[PackageName]/目录下，否则程序无法找到该文件。

源代码：InfaUtil.uploadDBFile。

```
//将apk中的raw/infa.db文件上传到Android系统文件夹中
public static void uploadDBFile(Context context) {
   SqliteDBHelper.updateDBFile(context, "/data/data/" + context.getPackageName()
+ "/databases/infa.db", R.raw.infa);
}
```

最终调用了 edu.sziit.common.sqlite.SqliteDBHelper.updateDBFile 方法：

源代码：edu.sziit.common.sqlite.SqliteDBHelper.updateDBFile。

```
public static void updateDBFile(Context context, String dataBasePath,
int dbRes) {//上传数据库文件到手机
   File dbfile=new File(dataBasePath);
   File dir=dbfile.getParentFile();
   if (!dir.exists()) {
      dir.mkdirs();
   }
   InputStream is=null;
   FileOutputStream fos=null;
   try {
      if (!dbfile.exists()) {
         //dbfile.createNewFile();
         is=context.getApplicationContext().getResources().openRawResource
(dbRes);
         fos=new FileOutputStream(dbfile);
         byte[] buff=new byte[is.available()];
         is.read(buff);
         fos.write(buff);
         Log.w(TAG, "Copying dbFile to "+dataBasePath+" successfully!");
      }
   } catch (Exception e) {
      String msg="Copying dbFile to "+dataBasePath+"failed!";
      Log.e(TAG, msg, e);
      throw new RuntimeException(msg);
   } finally {
      try {
         is.close();
      } catch (Exception e) {
      }
      try {
         fos.close();
      } catch (Exception e) {
      }
   }
}
```

然后启动程序进行测试，列表界面如图 8-7 所示。

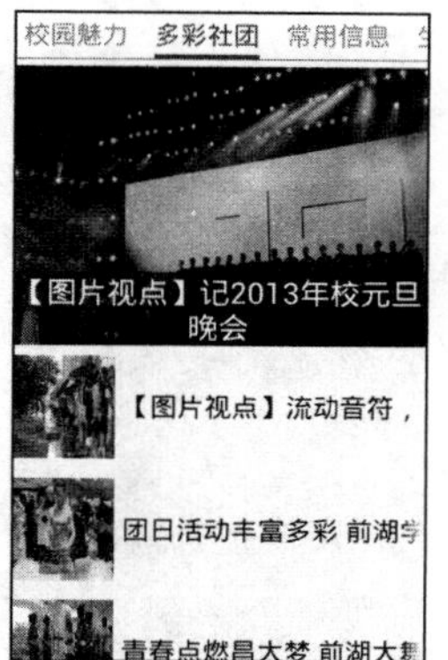

图 8-7　INFA-6 项目运行结果

8.4.8　添加详细信息页面

复制 INFA-6/INFA 项目到本地，导入 Eclipse 工作空间，并重命名项目为 INFA-7。本节将添加资讯详细信息页面和资讯图片信息全屏展示页面。创建资讯详细信息页面布局文件 res\layout\activity_zxdetail.xml，代码如下：

源代码：res\layout\activity_zxdetail.xml。

```
<?xml version="1.0" encoding="UTF-8"?>
<LinearLayout xmlns:android="http://schemas.android.com/apk/res/android"
    android:layout_width="fill_parent"
    android:layout_height="fill_parent"
    android:background="@color/white"
    android:orientation="vertical" >
    <RelativeLayout
        android:layout_width="fill_parent"
        android:layout_height="40dip"
        android:background="@color/list_navigator_text_select" >
        <ImageButton
            android:id="@+id/zxdetail_btn_back"
            android:layout_width="50dip"
            android:layout_height="wrap_content"
            android:layout_alignParentLeft="true"
            android:layout_centerVertical="true"
            android:layout_marginLeft="10dip"
            android:background="@drawable/btn_back_bg"
            android:visibility="visible" />
        <ImageButton
            android:id="@+id/zxdetail_btn_menu"
            android:layout_width="50dip"
            android:layout_height="wrap_content"
            android:layout_alignParentRight="true"
            android:layout_centerVertical="true"
            android:layout_marginRight="10dip"
            android:background="@drawable/btn_menu_bg"
            android:visibility="gone" />
        <TextView
            android:id="@+id/zxdetail_title"
            android:layout_width="wrap_content"
            android:layout_height="wrap_content"
            android:layout_centerInParent="true"
            android:layout_centerVertical="true"
            android:ellipsize="marquee"
            android:focusable="true"
            android:focusableInTouchMode="true"
            android:gravity="center_horizontal"
            android:singleLine="true"
            android:textColor="@color/white"
            android:textSize="@dimen/title_textSize" />
    </RelativeLayout>
    <ScrollView
        android:id="@+id/scrollView"
        android:layout_width="fill_parent"
        android:layout_height="wrap_content"
        android:layout_weight="1.0"
        android:fillViewport="true"
        android:scrollbars="vertical" >
        <LinearLayout
            android:id="@+id/lite_list"
            android:layout_width="fill_parent"
            android:layout_height="wrap_content"
```

```
        android:background="@color/white"
        android:orientation="vertical" >
        <FrameLayout
            android:id="@+id/framelayout"
            android:layout_width="fill_parent"
            android:layout_height="wrap_content" >
            <Gallery
                android:id="@+id/gallery_detail"
                android:layout_width="fill_parent"
                android:layout_height="220dp"
                android:persistentDrawingCache="animation"
                android:scrollbars="none"
                android:spacing="0dp" >
            </Gallery>
            <LinearLayout
                android:id="@+id/gallery_point_detail"
                android:layout_width="fill_parent"
                android:layout_height="wrap_content"
                android:layout_gravity="bottom"
                android:gravity="center" >
            </LinearLayout>
        </FrameLayout>
        <LinearLayout
            android:layout_width="fill_parent"
            android:layout_height="wrap_content"
            android:gravity="center_vertical"
            android:orientation="horizontal"
            android:padding="5dp" >
            <TextView
                android:id="@+id/zxdetail_content"
                android:layout_width="fill_parent"
                android:layout_height="wrap_content"
                android:layout_marginTop="3.0dip"
                android:layout_weight="1.0"
                android:lineSpacingMultiplier="1.5"
                android:text="@string/app_name"
                android:textAppearance="?android:textAppearanceMedium"
                android:textColor="@color/deep_gray"
                android:textSize="16.0sp" />
        </LinearLayout>
    </LinearLayout>
  </ScrollView>
</LinearLayout>
```

该布局上方显示一个标题栏，显示资讯的标题信息和返回按钮；接着是一个 Gallery，展示一组资讯图片；最下面是一个 TextView，展示资讯详细信息。

创建 src/edu/sziit/infa/adapter/DetailGalleryAdapter.java 类，该类是 Gallery 的适配器类，用于展示某个资讯的一组资讯图片，全部代码见 INFA-7\INFA\src\edu\sziit\infa\adapter\DetailGalleryAdapter.java 文件。

创建详细信息页面 Activity 类，该页面在主页面的列表项被点击时跳转。全部代码见 INFA-7\INFA\src\edu\sziit\infa\ui\ZxDetailActivity.java。

修改项目中 src/edu/sziit/infa/ui/MainActivity.java 中的 onListItemClick()方法，添加跳转到

详细信息页面的代码如下：

源代码：src/edu/sziit/infa/ui/MainActivity。

```
  private void onListItemClick (IFW_ZIXUN zx) {//响应列表点击事件
     Intent intent=new Intent(this, ZxDetailActivity.class);
     intent.putExtra("zixun", zx);
     this.startActivity(intent);
     this.finish();
  }
```

注意，此时 IFW_ZIXUN 对象需要作为参数传入 Intent 中，需要实现 Parcelable 接口。

在 AndroidManifest.xml 文件中配置该 Activity，在 application 结点下增加如下结点：

源代码：AndroidManifest.xml

```
<activity
    android:name=".ZxDetailActivity"
    android:screenOrientation="portrait" >
</activity>
```

运行项目，点击列表中某一项，进入详细信息页面如图 8-8 所示。

图 8-8　资讯详细信息页面

8.4.9　从服务器更新数据

复制 INFA-7/INFA 项目到本地，导入 Eclipse 工作空间，并重命名项目为 INFA-8。

本节讲从服务器更新数据到本地数据库。

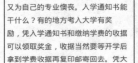

创建 edu.sziit.common.sao 包，包含服务器对象访问框架，实现了从服务器访问数据并封装为对象的框架，应用通过实现其定义的接口，可以实现具体对象的封装逻辑，包含的类如下：

edu.sziit.common.sao.IXmlMapper：定义了组装服务器返回的数据为对象的接口，代码参见：INFA-8\INFA\src\edu\sziit\common\sao\IXmlMapper.java。

edu.sziit.common.sao.SaoTool：实现服务器数据访问的框架，用户需实现 IXmlMapper 来组装服务器返回的数据为对象，程序流程如下：建立 HTTP 连接，获取 InputStream；创建 XmlPullParser 解析数据。代码如下：

源代码：edu.sziit.common.sao.SaoTool。

```
package edu.sziit.common.sao;
public class SaoTool {//服务器访问对象，从服务端获取数据
   /**
    * 从服务器获取对象数据
    * @param xmlPath http 请求的全路径 url.
    * @param mapper IXmlMapper 接口的实现，组装 xmlPath 对应请求的 XML 数据为对象
    * @return
    */
   public static <T> List<T> getList(String xmlPath, IXmlMapper<T> mapper) {
      InputStream inputStream=null;
      List<T> zixuns=new ArrayList<T>();
```

```
        HttpURLConnection urlConn=null;
        URL url=null;
        try {
            url=new URL(xmlPath);          //建立连接获取数据
            urlConn=(HttpURLConnection) url.openConnection();
            urlConn.setDoInput(true);
            urlConn.setDoOutput(true);
            urlConn.setUseCaches(false);
            urlConn.setRequestMethod("POST");
            urlConn.setRequestProperty("Content-Type", "text/xml;charset=utf-8");
            inputStream=urlConn.getInputStream();
            // 获取 XmlPullParser，解析数据
            XmlPullParserFactory factory=XmlPullParserFactory.newInstance();
            XmlPullParser xmlParser=factory.newPullParser();
            xmlParser.setInput(inputStream, "UTF-8");
            return mapper.parseList(xmlParser);
        } catch (Exception e) {
            e.printStackTrace();
        } finally {
            try {
                inputStream.close();
            } catch (Exception e) {
            }
            try {
                urlConn.disconnect();
            } catch (Exception e) {
            }
        }
        return zixuns;
    }
}
```

创建 edu.sziit.infa.sao.ZiXunMapper 类，实现 IXmlMapper 接口，将服务器获取的 XML 封装为 List<IFW_ZIXUN>对象。服务器返回的 XML 请参见 8.2.5 节。

创建异步任务 edu.sziit.infa.task.ZixunFromServer 类，从服务器获取数据，代码参见：INFA-8\INFA\src\edu\sziit\infa\task\ZixunFromServer.java。

修改项目中 src/edu/sziit/infa/ui/MainActivity.java 中的 onCreate()方法，添加下拉列表刷新的接口，实现代码如下（粗体部分），当列表下拉到顶时，继续下拉可以从服务器更新数据到本地数据库。

源代码：src/edu/sziit/infa/ui/MainActivity。

```
    public void onCreate(Bundle savedInstanceState) {        //初始化
        super.onCreate(savedInstanceState);
        requestWindowFeature(Window.FEATURE_NO_TITLE);        //不显示标题
        setContentView(R.layout.activity_main);
        listNavigatorRadioGroup=(RadioGroup)findViewById(R.id.list_ navigator_
radioGroup);
```

```
        listNavigatorLine=(ImageView) findViewById(R.id.list_navigator_ bottom_line);
        listNavigator=(HorizontalScrollView) findViewById(R.id.list_navigator);
        list=(MyListView) findViewById(R.id.list); //获取资讯列表
        list.setRefreshListener(new MyListView.RefreshListener() {
                                            // 设置列表更新的接口实现
            public void onRefresh() {
                new ZixunFromServer(MainActivity.this).execute(0);
            }
        });
        list.setOnItemClickListener(new OnItemClickListener () {
                                            //设置列表点击事件的接口实现
            public void onItemClick(AdapterView<?> arg0, View arg1, int arg2,long arg3){
                IFW_ZIXUN zx=ListAdapter.getZixun(arg1);
                onListItemClick (zx);
            }
        });
        new FenleiFromDB(this).execute(0);    //从数据库获取资讯分类数据
    }
```


该实现中启动了一个 ZixunFromServer 任务，从服务器获取数据并入本地数据库，获取图片存储在 SD 卡上，并回调 MainActivity 的 afterZixunRefresh()方法，如果数据刷新成功，则启动 ZixunFromDB 任务，从本地数据库更新列表视图数据。

由于程序需要访问外部网络，并且存储图片到 SD 卡上，所以需要在项目 AndroidManifest.xml 文件中配置相关权限如下：

源代码：AndroidManifest.xml。

```
<!-- 访问网络的权限 -->
    <uses-permission android:name="android.permission.INTERNET" />
    <uses-permission android:name="android.permission.ACCESS_NETWORK_STATE" />
    <!-- 写 SD 卡权限 -->
    <uses-permission android:name="android.permission.WRITE_EXTERNAL_ STORAGE" />
```

8.4.10 搭建服务器环境进行测试

解压 INFA-8/IFW.zip 文件到本地后：单击 start-derby.bat 启动数据库，如图 8-9 所示。

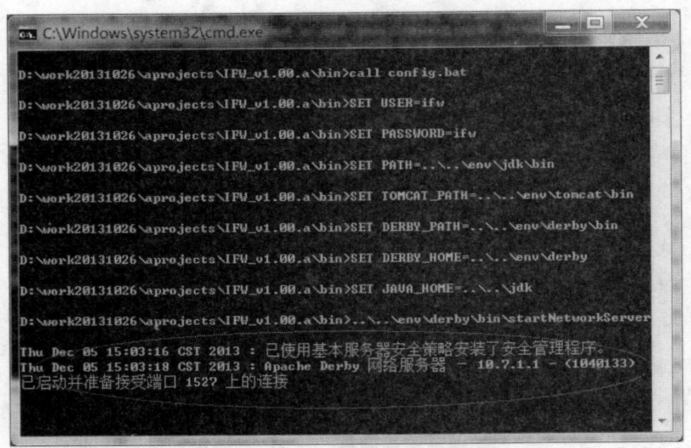

图 8-9　数据库启动成功

数据库启动后，出现图 8-9 中圈中的文字时，表示启动成功，然后单击 start-tomcat.bat 启动 Tomcat，如图 8-10 所示。

图 8-10　Tomcat 启动成功

Tomcat 启动后，出现图 8-10 中圈中的文字时，表示启动成功，此时打开浏览器访问如下地址（localhost 换为实际服务器 IP 地址即可）：http://localhost:8080/IFW/，如图 8-11 所示。

图 8-11　校园通服务器登录页面

输入登录用户名和密码：ifwuser/ifwuser 后进入系统，如图 8-12 所示。

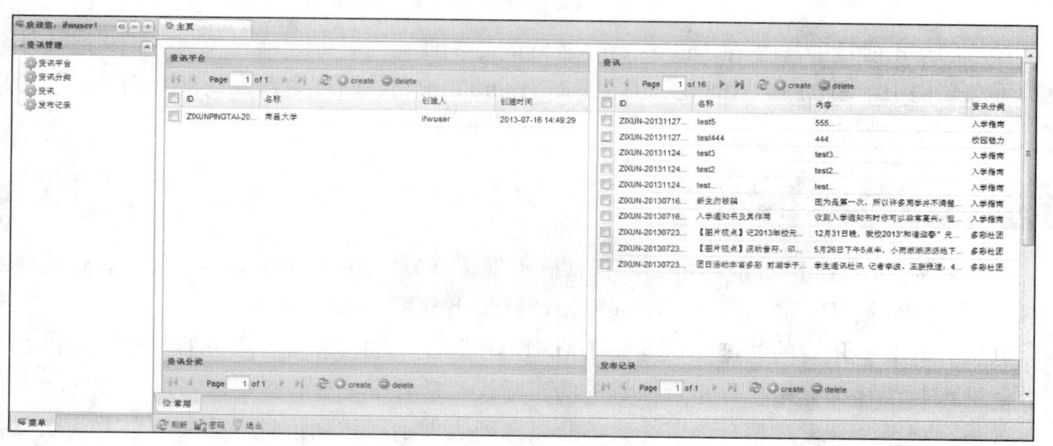

图 8-12　校园通服务器主页面

单击“资讯”菜单，并单击 create 按钮，如图 8-13 所示。

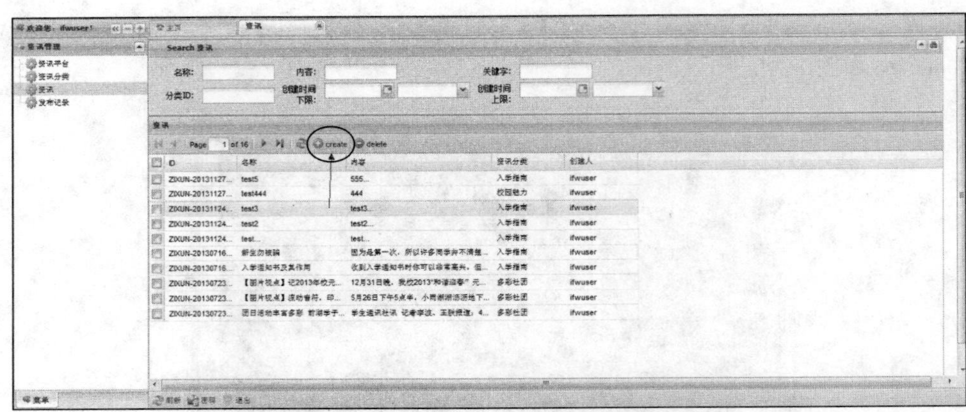

图 8-13　校园通服务器资讯管理页面

名称、内容、小图字段必须填写，图 1～3 中至少上传一个，资讯分类选择“入学指南”，然后单击 submit 按钮，如图 8-14 所示。

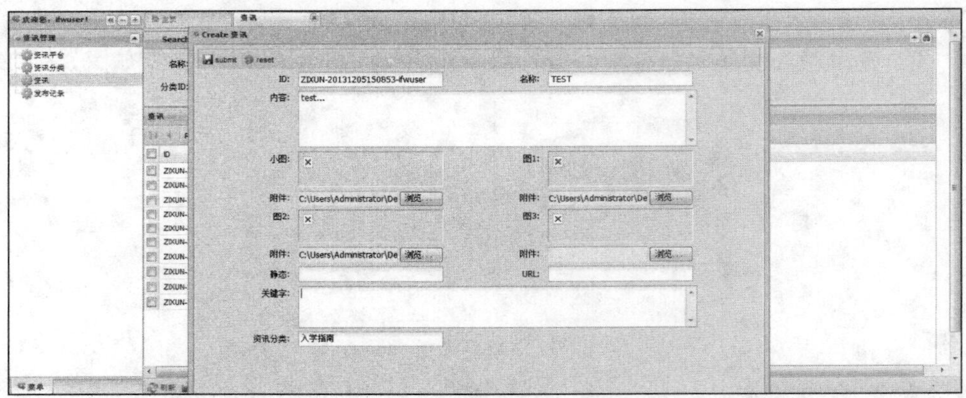

图 8-14　校园通服务器创建资讯页面

在 Android 项目中修改 res/values/strings.xml 文件的 server_url 参数，将其中 IP 地址部分改为服务器的 IP，例如 192.168.43.203，运行 Android 项目，即可更新最新的数据到“入学指南”栏目的下拉列表中。

8.5　本章小结

本章首先基于 Android 应用框架项目基础上继续开发，完成校园通项目 Android 侧软件的需求分析、总体设计、概要设计、数据库设计及编码测试工作。

其次，对其中用到的关键知识点如 DAO 设计模式、XML 解析、服务器数据访问以及异步任务等进行了详细介绍，并结合项目中的实际代码进行了分析，并给出了示例。

最后，把整个开发过程分为三个大的阶段（列表展示、详细信息页面展示以及获取服务器数据），每个阶段加入新的功能进行迭代开发，具体请参见项目目录：INFA-6/INFA、INFA-7/INFA 及 INFA-8/INFA。

第9章 基于 Android 智能仓储系统项目

学习目标：

- 界面的设计和实现；
- 页面与页面之间数据的交互；
- 数据库的设计和实现；
- 视频数据处理。

9.1 项目概述

物联网智能仓储系统是基于 RFID 技术、Zigbee 技术、WI-FI 技术等的可用于真实项目的系统。整个系统具有五大子系统：环境监控系统、安防报警系统、入库系统、出库系统、查询系统。

无论何时何地，都可以通过手持设备或者计算机控制仓储环境状况、货物入库记录、货物出库记录等功能，同时还支持防盗报警、火灾报警等功能。集成了目前智能仓储中众多声光电因素，能够很好地展示现代仓储管理的安全化和智能化。

智能仓储实训台高度结合了物联网工程技术与行业体系架构，还原了行业真实环境，将感知层、网络层和应用层三个区域分开、区域清晰；预留扩展接口，方便用户二次开发调试；烧写接口集中管控方便烧写；智能断电系统，过电流可自动切断电源。

9.2 项目设计

9.2.1 项目总体功能需求

1. 环境监控系统

在手持设备或平板电脑上能查看到：温度、湿度、光照度、空气质量等传感器的状况。

2. 安防报警系统

（1）防盗报警：当仓库里没有人，系统检测到有人入侵时，启动报警器，同时，手持设备或平板电脑能显示报警信息。

（2）防灾报警：当仓库里有烟雾异常时，启动报警器，打开通风设备，同时，手持设备或平板电脑能显示报警信息。

3. 入库系统

（1）通过 RFID 设备对货物进行自动入库，货物的入库信息经智能网关定时上传到系统服务器中。

（2）在入库系统中通过选择不同的时间段可以直观地查看每天、每周、每月货物入库信息的曲线图。

（3）在入库系统中通过选择具体的货物，可以查看该货物的入库详情：货物名称、货物产地、货物入库人员名称、货物入库时间。

4. **出库系统**

（1）通过 RFID 设备对货物进行自动出库，货物的出库信息经智能网关定时上传到系统服务器中。

（2）在出库系统中通过选择不同的时间段可以直观地查看每天、每周、每月货物出库信息的曲线图。

（3）在出库系统中通过选择具体的货物，可以查看该货物的出库详情：货物名称、货物销售地、货物出库人员名称、货物出库时间。

5. **查询系统**

在查询系统中通过输入具体的货物，可以查看该货物的总库存数据、入库总数量、入库价格、出库总数量、出库价格。

9.2.2 项目总体设计

现在全球在进行信息化建设，而信息化建设最终能否落地，移动终端上的应用开发将起到决定性作用，Android 是目前市场占有率最高的移动设备操作系统，同时它的操作系统是开源的，这样有利于其上应用程序的开发。因此，本系统的应用程序选择在 Android 操作系统上进行开发。本项目整体设计框架如图 9-1 所示。

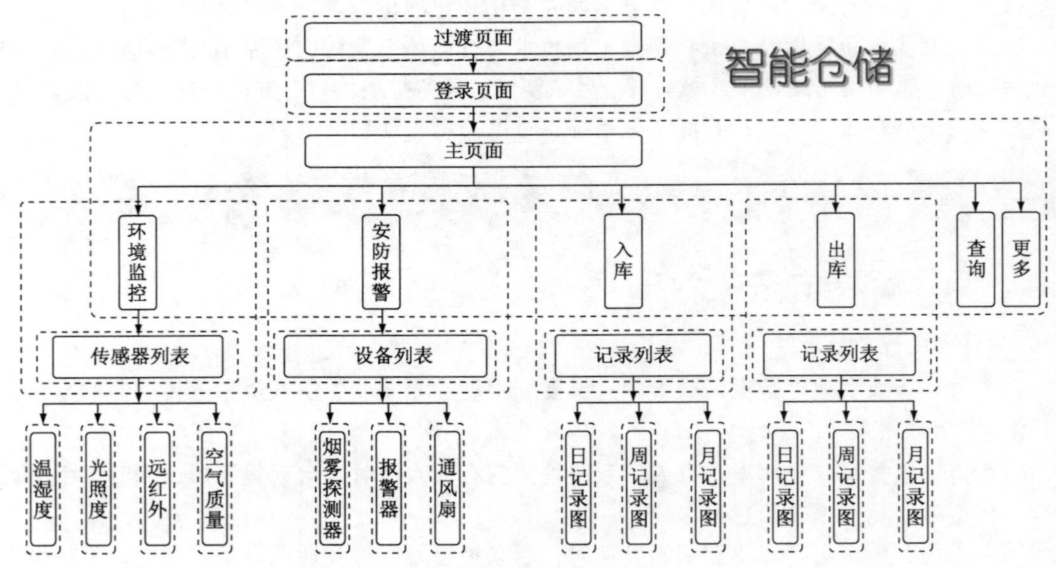

图 9-1　项目整体设计框架图

从安全性和便利性方面考虑，本系统将采用云与端的方式进行开发。

本系统采用常规的用户与密码的方式，登录进入系统，在系统主页面中有六个采用扁平化设计的功能选项：环境监控、安防报警、物品入库、物品出库、物品查询设置和更多。

在“环境监控”中包括：温湿度、光照度、空气质量、远红外和“+”；“+”的功能主要是考虑功能的扩展，用户可以根据仓库的情况添加新的探测器，同时也可以对探测器进行删除操作。在界面中查看到简单的探测器状况信息。

在“安防报警”中包括：烟雾探测器、报警器、通风扇和“+”；“+”的功能主要是考虑功能的扩展，用户可以根据实际情况添加新的设备，同时也可以对设备进行删除操作。在界面中查看到简单的设备状况信息。对于可控制的设备，点击具体的设备后，可以查看设备具体信息，并可以对设备进行控制。

在“入库”中包括：日记录图、周记录图、月记录图和“+”；“+”的功能主要是考虑功能的扩展。

在“出库”中包括：日记录图、周记录图、月记录图和“+”；“+”的功能主要是考虑功能的扩展。

“查询”中主要用于货物的查询。

“更多”选项用于后期功能的扩展。

9.3 必备的技术和知识点

本章必备的技术和知识点包括：

（1）界面编程与视图组件。

（2）布局管理器。

（3）TextView 及其子类。

（4）ImageView 及其子类。

（5）AdapterView 及其子类。

（6）ViewAnimator 及其子类。

（7）对话框和菜单。

（8）基于监听的事件处理。

（9）基于回调的事件处理。

（10）响应系统设置的事件。

（11）Handle 消息传递机制。

（12）基于 TCP 的网络通信。

（13）使用 URL 访问网络资源。

（14）使用 WebServer 进行网络编程。

（15）数据存储与访问。

（16）嵌入式关系型 SQLite 数据库存储数据。

（17）视频采集。

9.4 项目实施

9.4.1 登录页面

实现账号和密码输入文本框、实现登录按键功能、实现保存密码和自动登录功能。页面

布局如图 9-2 所示。

图 9-2　登录页面图

当用户输入账号和密码，点击登录按键时，系统读取账号和密码文本框信息，并把账号和密码信息经网络传输给智能网关，与智能网关中的数据库中的账号和密码数据信息进行比对，如果账号和密码同时存在于智能网关中的数据库中，则允许用户登录系统。如果账号或密码不正确，则根据返回标记码的不同，提示“账号不存在”或“密码不正确”。

当用户选中“保存密码”时，系统会把账号和密码保存在系统相关的配置文件中，当下次再登录系统时，则不需要再次输入账号和密码。如果用户只选中“自动登录”时，则系统会保存账号和密码，下次自动登录系统。

程序的界面布局代码：

程序清单：CH09\WareHousing\src\com\warehousing \smarthome \activity\ LoginActivity.java。

```
package com.zigcloud.warehousing.activity;
public class LoginActivity extends BaseActivity{ /** * 用户登录* */
  public void onCreate(Bundle savedInstanceState) {
     super.onCreate(savedInstanceState);
     setContentView(R.layout.activity_login);
     MainApplication.getInstance().addActivity(this);
     findViewById(R.id.btn_login).setOnClickListener(new View.OnClickListener() {
        public void onClick(View v) {
           startActivity(new Intent(LoginActivity.this,MainActivity.class));
        }
     });
  }
  public void onBackPressed() {
     super.onBackPressed();
  }
  public boolean onKeyDown(int keyCode, KeyEvent event) {
     if(keyCode==KeyEvent.KEYCODE_BACK)
     {
         exitBy2Click();
     }
     return false;
  }
  private static Boolean isExit=false;
  private void exitBy2Click() {/*** 双击后退按钮退出* */
     Timer tExit=null;
     if (isExit==false) {
        isExit=true;
        Toast.makeText(this, getResources().getString(R.string.exit_dialog_
message),Toast.LENGTH_SHORT).show();
        tExit=new Timer();
        tExit.schedule(new TimerTask() {
           public void run() {
              isExit=false;
           }
```

```
            }, 2000);
        } else {
            MainApplication.getInstance().exit();
        }
    }
}
```

9.4.2 主页面

图 9-3 主页面图

实现“返回”和六个示图功能。页面布局如图 9-3 所示。

“返回”功能的作用返回上一级页面；“环境监控”功能的作用是查看当前仓库中的传感器设备信息。其他的五张示图功能类似。程序的界面布局代码清单。

程序清单：CH09\WareHousing\res\layout\activity_main.xml。

```
<?xml version="1.0" encoding="utf-8"?>
<RelativeLayout xmlns:android="http://schemas. android.com/apk/res/android"
    android:layout_width="fill_parent"
    android:layout_height="fill_parent"
    android:background="@drawable/main_default_ bg" >
    <include android:id="@id/bt_createtask_title_ layout"
        layout="@layout/activity_common_title_ bar" />
    <RelativeLayout
        android:id="@+id/RelativeLayout1"
        android:layout_width="match_parent"
        android:layout_height="match_parent"
        android:layout_below="@id/bt_createtask_title_layout"
        android:layout_centerHorizontal="true"
        android:layout_marginLeft="15dp"
        android:layout_marginRight="15dp"
        android:orientation="vertical"
        android:visibility="visible" >
        <LinearLayout
            android:id="@+id/LinearLayout1"
            android:layout_width="match_parent"
            android:layout_height="match_parent"
            android:layout_alignParentTop="true"
            android:orientation="vertical" >
            <GridView
              android:id="@+id/gdv_main"
              android:layout_width="match_parent"
              android:layout_height="wrap_content"
              android:listSelector="@drawable/selector_list"
              android:horizontalSpacing="10dp"
              android:numColumns="2"
              android:verticalSpacing="10dp" >
          </GridView>
            <include
                android:id="@+id/view_loading_error"
                android:layout_width="match_parent"
                android:layout_height="match_parent"
                layout="@layout/view_loading_error"
```

```
            android:visibility="gone" />
        <include
            android:id="@+id/view_loading"
            android:layout_width="match_parent"
            android:layout_height="match_parent"
            layout="@layout/view_loading"
            android:visibility="gone" />
      </LinearLayout>
   </RelativeLayout>
</RelativeLayout>
```

程序清单：CH09\ WareHousing\src\com\zigcloud\warehousing \activity\MainActivity.java。

```
package com.zigcloud.warehousing.activity;
public class MainActivity extends BaseActivity{  /** * 主界面* */
   private TextView titlebar_left;
   private TextView titlebar_title;
   public void onCreate(Bundle savedInstanceState) {
      super.onCreate(savedInstanceState);
      setContentView(R.layout.activity_main);
      MainApplication.getInstance().addActivity(this);
      titlebar_left=(TextView) findViewById(R.id.titlebar_left);
      titlebar_left.setOnClickListener(new View.OnClickListener() {
         public void onClick(View v) {
            onBackPressed();
         }
      });
      titlebar_title=(TextView) findViewById(R.id.titlebar_title);
      titlebar_title.setText(R.string.app_name);
     GridView gridView=(GridView) findViewById(R.id.gdv_main);
     ArrayList<HashMap<String,Object>>lstImageItem=new ArrayList <HashMap
<String, Object>>();
     HashMap<String, Object> map=new HashMap<String,Object>();
      map.put("ItemImage", R.drawable.icon_environment);
      map.put("ItemText", "环境监控");
      lstImageItem.add(map);
      map=new HashMap<String, Object>();
      map.put("ItemImage", R.drawable.icon_alarm);
      map.put("ItemText", "安防报警");
      lstImageItem.add(map);
      map=new HashMap<String, Object>();
      map.put("ItemImage", R.drawable.icon_import);
      map.put("ItemText", "物品入库");
      lstImageItem.add(map);
      map=new HashMap<String, Object>();
      map.put("ItemImage", R.drawable.icon_outport);
      map.put("ItemText", "物品出库");
      lstImageItem.add(map);
      map=new HashMap<String, Object>();
      map.put("ItemImage", R.drawable.icon_search);
      map.put("ItemText", "物品查询");
      lstImageItem.add(map);
      map=new HashMap<String, Object>();
      map.put("ItemImage", R.drawable.icon_setting);
```

```
        map.put("ItemText", "设置");
        lstImageItem.add(map);
        map=new HashMap<String, Object>();
        map.put("ItemImage", R.drawable.icon_add);
        map.put("ItemText", "预留");
        lstImageItem.add(map);
        map=new HashMap<String, Object>();
        map.put("ItemImage", R.drawable.icon_add);
        map.put("ItemText", "预留");
        lstImageItem.add(map);
      SimpleAdapter saImageItems=new SimpleAdapter(this, lstImageItem, R.
layout.view_squared_item, new String[] {"ItemImage","ItemText"}, new int[] {R.
id.itemImage, R.id.itemText});
        gridView.setAdapter(saImageItems);
        gridView.setOnItemClickListener(new OnItemClickListener() {
           public void onItemClick(AdapterView<?> arg0, View arg1, int arg2,long arg3) {
              switch(arg2){
                case 0:
                   startActivity(new Intent ntent(MainActivity.this,Environment
MonitorActivity.class));
                   break;
                case 1:
                   break;
                 case 2:
                   startActivity(new Intent(MainActivity.this,GoodsImportActivity.
class));
                   break;
                 case 3:
                   startActivity(new Intent(MainActivity.this,GoodsOutportActivity.
class));
                   break;
                 case 4:
                   startActivity(new Intent(MainActivity.this,GoodsSearchActivity.
class));
                   break;
              }
           }
        });
          gridView.setSelector(new ColorDrawable(Color.TRANSPARENT));
      }
    public boolean onKeyDown(int keyCode, KeyEvent event) {
        if(keyCode==KeyEvent.KEYCODE_BACK)
        {
            exitBy2Click();
        }
        return false;
    }
    private static Boolean isExit=false;
    private void exitBy2Click() {
        Timer tExit=null;
        if(isExit==false) {
           isExit=true;
```

```
            Toast.makeText(this,getResources().getString(R.string.exit_dialog_
message), Toast.LENGTH_SHORT).show();
            tExit=new Timer();
            tExit.schedule(new TimerTask() {
               public void run() {
                  isExit=false;
               }
            }, 2000);
         } else {
            MainApplication.getInstance().exit();
         }
      }
   }
```

9.4.3　环境监控页面

实现查看仓库中各个区域的传感器信息，如火焰传感器、温度传感器、湿度传感器等环境信息。环境监控页面如图 9-4 所示。

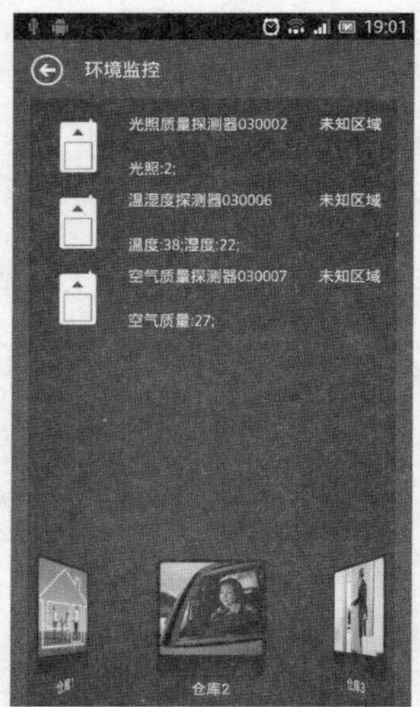

图 9-4　环境监测页面图

程序界面布局代码：CH09\WareHousing\res\layout\ activity_environmentmonitor.xml。

```
<?xml version="1.0" encoding="utf-8"?>
<RelativeLayout xmlns:android="http://schemas.android.com/apk/res/android"
    android:layout_width="fill_parent"
    android:layout_height="fill_parent"
    android:background="@drawable/main_default_bg" >
    <include
        android:id="@id/bt_createtask_title_layout"
        layout="@layout/activity_common_title_bar" />
    <RelativeLayout
        android:id="@+id/RelativeLayout1"
        android:layout_width="match_parent"
        android:layout_height="match_parent"
        android:layout_below="@id/bt_createtask_title_layout"
        android:layout_centerHorizontal="true"
        android:layout_marginLeft="15dp"
        android:layout_marginRight="15dp"
        android:background="@drawable/content_default_bg"
        android:orientation="vertical"
        android:visibility="visible" >
        <com.zigcloud.warehousing.widget.Gallery3D
            android:id="@+id/gal_rooms"
            android:layout_width="fill_parent"
            android:layout_height="wrap_content"
            android:layout_alignParentBottom="true"
            android:layout_alignParentLeft="true"
            android:spacing="30dp"
            android:unselectedAlpha="128" />
        <LinearLayout
            android:id="@+id/LinearLayout1"
            android:layout_width="match_parent"
```

```
            android:layout_height="match_parent"
            android:layout_above="@+id/gal_rooms"
            android:layout_alignParentTop="true"
            android:orientation="vertical" >
            <include
               android:id="@+id/view_loading_error"
               android:layout_width="match_parent"
               android:layout_height="match_parent"
               layout="@layout/view_loading_error"
               android:visibility="gone" />
            <include
               android:id="@+id/view_loading"
               android:layout_width="match_parent"
               android:layout_height="match_parent"
               layout="@layout/view_loading"
               android:visibility="gone" />
            <GridView
               android:id="@+id/gdv_equipments"
               android:layout_width="match_parent"
               android:layout_height="wrap_content"
               android:numColumns="1"
               android:padding="10dp"
               android:verticalSpacing="10dp"
               android:visibility="visible" >
            </GridView>
         </LinearLayout>
      </RelativeLayout>
   </RelativeLayout>
```

程序清单：CH09\WareHousing\src\com\zigcloud\warehousing\activity\ EnvironmentMonitorActivity.java。

```
   package com.zigcloud.warehousing.activity;
   @SuppressLint("HandlerLeak")
   public class EnvironmentMonitorActivity extends BaseActivity{ /*** 环境监控**/
      private TextView titlebar_left;
      private TextView titlebar_title;
      private GridView equipmentsGridView;
      private  ArrayList<HashMap<String,  Object>>  equipmentsArrayList=new
ArrayList<HashMap<String, Object>>();
      private HashMap<String,Object> equipmentsHashMap=new HashMap<String, Object>();
      private SimpleAdapter equipmentsAdapter;
      /*** 获取设备列表**/
      private EquipmentListHttpRequestTask equipmentListHttpRequestTask=new
EquipmentListHttpRequestTask();
      public void onCreate(Bundle savedInstanceState){
         super.onCreate(savedInstanceState);
         setContentView(R.layout.activity_environmentmonitor);
         MainApplication.getInstance().addActivity(this);
         titlebar_left=(TextView) findViewById(R.id.titlebar_left);
         titlebar_left.setOnClickListener(new View.OnClickListener() {
            public void onClick(View v) {
               onBackPressed();
```

```
            }
        });
        titlebar_title=(TextView) findViewById(R.id.titlebar_title);
        titlebar_title.setText(R.string.title_environment_monitor);
        initialRoomsGallery();
        initialEquipmentsGridView();
    }
    private void initialEquipmentsGridView(){/*** 初始化设备信息列表* */
        equipmentsGridView=(GridView) findViewById(R.id.gdv_equipments);
        equipmentsAdapter=new SimpleAdapter(getApplicationContext(),
equipmentsArrayList,R.layout.activity_equipment_list_item,new String[] {"ItemImage",
"ItemNodeName","ItemDataValue","ItemArea"},new int[] {R.id.img_image,R.id.tv_name,
R.id.tv_datavalue,R.id. tv_area});
        equipmentsGridView.setAdapter(equipmentsAdapter);
        equipmentsGridView.setOnItemClickListener(new OnItemClickListener() {
            @SuppressWarnings("unchecked")
            public void onItemClick(AdapterView<?> arg0, View arg1, int arg2,
    long arg3) {
                HashMap<String, Object> item=(HashMap<String, Object>) arg0.
getItemAtPosition(arg2);
                String nodeId=item.get("ItemNodeId")==null ?null:item.get
("ItemNodeId").toString();
                String nodeName=item.get("ItemNodeName")==null ?null:item.
get("ItemNodeName").toString();
                String nodeTypeId=item.get("ItemNodeTypeId")==null ?null:item.get
("ItemNodeTypeId").toString();
                String nodeTypeName=item.get("ItemNodeTypeName")==null ?null:item.get
("ItemNodeTypeName").toString();
                String dataValueString=item.get("ItemDataValueString")==null ?null:item.
get("ItemDataValueString").toString();
                String updateTimeString=item.get("ItemUpdateTimeString")==null ?null:item.
get("ItemUpdateTimeString").toString();
                Intent intent=new Intent(getApplicationContext(),Equipment Activity.
class);
                intent.putExtra("nodeId", nodeId);
                intent.putExtra("nodeName", nodeName);
                intent.putExtra("nodeTypeId", nodeTypeId);
                intent.putExtra("nodeTypeName", nodeTypeName);
                intent.putExtra("dataValueString", dataValueString);
                intent.putExtra("updateTimeString",updateTimeString);
                startActivity(intent);
            }
        });
        equipmentListHttpRequestTask.execute();
    }
    private EquipmentDAO mEquipmentDAO=new EquipmentDAO();/*** 设备业务控制类**/
    private Handler mHandler=new Handler(){
        public void handleMessage(Message msg) {
            super.handleMessage(msg);
            switch(msg.what){
                case 10001:                    //正在加载
                    findViewById(R.id.gdv_equipments).setVisibility(8);
                    findViewById(R.id.view_loading).setVisibility(0);
```

```
                findViewById(R.id.view_loading_error).setVisibility(8);
                break;
            case 10002:                      //加载成功
                findViewById(R.id.gdv_equipments).setVisibility(0);
                findViewById(R.id.view_loading).setVisibility(8);
                findViewById(R.id.view_loading_error).setVisibility(8);
                break;
            case 10003:                      //加载失败
                findViewById(R.id.gdv_equipments).setVisibility(8);
                findViewById(R.id.view_loading).setVisibility(8);
                findViewById(R.id.view_loading_error).setVisibility(0);
                break;
            }
        }
    };
    /*** 获取equipment列表* */
    private class EquipmentListHttpRequestTask extends AsyncTask<String,
Integer, List<EquipmentJson>>{protected void onPreExecute() {
            super.onPreExecute();
            if(!(equipmentsAdapter!=null&&equipmentsAdapter.getCount()>0)){
                mHandler.sendEmptyMessage(10001);
            }
        }
        protected List<EquipmentJson> doInBackground(String... params) {
            return mEquipmentDAO.getAll();
        }
        protected void onProgressUpdate(Integer... values) {
            super.onProgressUpdate(values);
        }
        protected void onPostExecute(List<EquipmentJson> result) {
            if(result!=null){
                EquipmentJson equipmentEntity=null;
                BaseEquipmentEntity equipment=null;
                for(int i=0;i<result.size();i++){
                    equipmentEntity= result.get(i);
                    equipment= BaseEquipmentEntity.parse(equipmentEntity);
                    if(equipment!=null){
                        equipmentsHashMap=new HashMap<String, Object>();
                        equipmentsHashMap.put("ItemImage", equipment.getIconRes());
                        equipmentsHashMap.put("ItemNodeName", equipment.getName());
                        equipmentsHashMap.put("ItemDataValue", equipment.getData
ValueString());
                        equipmentsHashMap.put("ItemArea", "未知区域");
                        equipmentsHashMap.put("ItemNodeId",equipment.getNodeId());
                        equipmentsHashMap.put("ItemNodeTypeId",equipment.
getNodeTypeId());
                        equipmentsHashMap.put("ItemNodeTypeName",equipment.
getTypeName());
                        equipmentsHashMap.put("ItemDataValueString",
equipment.getDataValueString());
                        equipmentsHashMap.put("ItemUpdateTimeString",equipment.
getUpdateTimeString());
                        equipmentsArrayList.add(equipmentsHashMap);
```

```
                }
            }
            equipmentsAdapter.notifyDataSetChanged();
            mHandler.sendEmptyMessage(10002);
        }
        else{
            mHandler.sendEmptyMessage(10003);
        }
    }
};
private void initialRoomsGallery(){
    Gallery3D  gallery=(Gallery3D) findViewById(R.id.gal_rooms);
    ArrayList<HashMap<String,  Object>>  lstImageItem=new  ArrayList
<HashMap<String, Object>>();
    HashMap<String, Object> map=new HashMap<String, Object>();
    map.put("ItemImage", R.drawable.scene_gallery_0);
    map.put("ItemText", "仓库1");
    map.put("ItemContent", "仓库1");
    lstImageItem.add(map);
    map=new HashMap<String, Object>();
    map.put("ItemImage", R.drawable.scene_gallery_1);
    map.put("ItemText", "仓库2");
    map.put("ItemContent", "仓库2");
    lstImageItem.add(map);
    map=new HashMap<String, Object>();
    map.put("ItemImage", R.drawable.scene_gallery_2);
    map.put("ItemText", "仓库3");
    map.put("ItemContent", "仓库3");
    lstImageItem.add(map);
    map=new HashMap<String, Object>();
    map.put("ItemImage", R.drawable.scene_gallery_3);
    map.put("ItemText", "仓库4");
    map.put("ItemContent", "仓库4");
    lstImageItem.add(map);
    SimpleAdapter saImageItems=new SimpleAdapter(this, lstImageItem,R.
layout.view_grallery3d_item,new String[] {"ItemImage","ItemText","ItemContent"},
new int[] {R.id.itemImage,R.id.itemText,R.id.itemContent});
    gallery.setFadingEdgeLength(0);
    gallery.setAdapter(saImageItems);
    gallery.setOnItemSelectedListener(new OnItemSelectedListener() {
        public void onItemSelected(AdapterView<?> parent, View view, int
position, long id) {
            Toast.makeText(RoomListActivity.this, "img " + (position+1) +
" selected", Toast.LENGTH_SHORT).show();
        }
        public void onNothingSelected(AdapterView<?> parent) {
        }
    });
    gallery.setOnItemClickListener(new OnItemClickListener() {
        public void onItemClick(AdapterView<?> parent, View view, int
position, long id) {
            Toast.makeText(RoomListActivity.this,  "img  "+(position+1)+"
selected",Toast.LENGTH_SHORT).show();
```

```
            }
        });
    }
}
```

9.4.4 物品入库页面

图 9-5 物品入库页面

实现货物入库的功能。页面布局如图 9-5 所示。

程序的界面布局代码如下：CH09\WareHousing\res\layout\activity_goods_import.xml。

```
<?xml version="1.0" encoding="utf-8"?>
<RelativeLayout xmlns:android=
"http://schemas.android.com/apk/res/android"
    android:layout_width="fill_parent"
    android:layout_height="fill_parent"
    android:background="@drawable/main_default_bg" >
    <include
        android:id="@+id/bt_createtask_title_layout"
        layout="@layout/activity_common_title_bar" />
    <RelativeLayout
        android:id="@+id/RelativeLayout1"
        android:layout_width="match_parent"
        android:layout_height="match_parent"
        android:layout_below="@id/bt_createtask_title_layout"
        android:layout_centerHorizontal="true"
        android:layout_marginLeft="15dp"
        android:layout_marginRight="15dp"
        android:background="@drawable/content_default_bg"
        android:orientation="vertical"
        android:visibility="visible" >
        <TableLayout
            android:layout_width="wrap_content"
            android:layout_height="wrap_content"
            android:layout_alignParentBottom="true"
            android:layout_alignParentLeft="true"
            android:layout_alignParentRight="true"
            android:layout_alignParentTop="true" >
            <TableRow
                android:id="@+id/tableRow1"
                android:layout_width="wrap_content"
                android:layout_height="wrap_content" >
                <TextView
                    android:id="@+id/textView1"
                    android:layout_width="wrap_content"
                    android:layout_height="wrap_content"
                    android:text="@string/goods_cardid" />
                <EditText
```

```
            android:id="@+id/edt_goods_cardId"
            android:layout_width="wrap_content"
            android:layout_height="wrap_content"
            android:ems="10" >
            <requestFocus />
        </EditText>
    </TableRow>
    <TableRow
        android:id="@+id/tableRow2"
        android:layout_width="wrap_content"
        android:layout_height="wrap_content" >
        <TextView
            android:id="@+id/textView2"
            android:layout_width="wrap_content"
            android:layout_height="wrap_content"
            android:text="@string/goods_name" />
        <EditText
            android:id="@+id/edt_goods_name"
            android:layout_width="wrap_content"
            android:layout_height="wrap_content"
            android:ems="10" />
    </TableRow>
    <TableRow
        android:id="@+id/tableRow3"
        android:layout_width="wrap_content"
        android:layout_height="wrap_content" >
        <TextView
            android:id="@+id/textView3"
            android:layout_width="wrap_content"
            android:layout_height="wrap_content"
            android:text="@string/goods_address"/>
        <EditText
            android:id="@+id/edt_goods_address"
            android:layout_width="wrap_content"
            android:layout_height="wrap_content"
            android:ems="10" />
    </TableRow>
    <TableRow
        android:id="@+id/tableRow4"
        android:layout_width="wrap_content"
        android:layout_height="wrap_content" >
        <TextView
            android:id="@+id/textView4"
            android:layout_width="wrap_content"
            android:layout_height="wrap_content"
            android:text="@string/operate_option" />
```

```
            <Button
                android:id="@+id/btn_ok"
                android:layout_width="wrap_content"
                android:layout_height="wrap_content"
                android:text="@string/ok" />
        </TableRow>
    </TableLayout>
  </RelativeLayout>
</RelativeLayout>
```

程序清单：CH09\WareHousing\src\com\zigcloud\warehousing\activity\GoodsImportActivity.java。

```
package com.zigcloud.warehousing.activity;
public class GoodsImportActivity extends BaseActivity{/*** 物品入库**/
  private TextView titlebar_left;
  private TextView titlebar_title;
  private GoodsDAO mGoodsDAO=new GoodsDAO();/*** 物品操作业务类**/
  private GoodsImportTask mGoodsImportTask=new GoodsImportTask();
  /*** 物品入库任务类**/
  protected void onCreate(Bundle savedInstanceState) {
     super.onCreate(savedInstanceState);
     setContentView(R.layout.activity_goods_import);
     MainApplication.getInstance().addActivity(this);
     titlebar_left=(TextView) findViewById(R.id.titlebar_left);
     titlebar_left.setOnClickListener(new View.OnClickListener() {
        public void onClick(View v) {
           onBackPressed();
        }
     });
     titlebar_title=(TextView) findViewById(R.id.titlebar_title);
     titlebar_title.setText(R.string.title_goods_import);
     /*** 商品入库按钮事件* */
     findViewById(R.id.btn_ok).setOnClickListener(new OnClickListener() {
        public void onClick(View v) {
           EditText edt_goods_cardId=(EditText)findViewById(R.id.edt_goods_cardId);
           EditText edt_goods_name=(EditText)findViewById(R.id.edt_goods_ name);
           EditText edt_goods_address=(EditText)findViewById(R.id.edt_goods_ address);
           mGoodsImportTask=new GoodsImportTask();
           mGoodsImportTask.execute(edt_goods_cardId.getText().toString(),
edt_goods_name.getText().toString(),edt_goods_address.getText().toString(),"admin");
        }
     });
  }
  /*** 物品入库任务类* */
  private class GoodsImportTask extends AsyncTask<String,Integer,ControlResultJson>{
     protected ControlResultJson doInBackground(String... params) {
        if(params!=null&&params.length>3){
           String cardNum=params[0];
           String dataName=params[1];
           String originPlace=params[2];
           String staffName=params[3];
```

```
            if(cardNum!=null&&dataName!=null&&originPlace!=null&&staff
Name!=null)
                return mGoodsDAO.goodsImport(cardNum, dataName, originPlace,
staffName);
            }
            return null;
        }
        protected void onPostExecute(ControlResultJson result) {
            super.onPostExecute(result);
            String resStr=null;
            if(result!=null){
                resStr=result.flag==0?"添加成功! ":"添加失败";
            }
            else{
                resStr="添加失败";
            }
            Toast.makeText(GoodsImportActivity.this,resStr,Toast.LENGTH_ SHORT).show();
        }
        protected void onPreExecute() {
            super.onPreExecute();
        }
        protected void onProgressUpdate(Integer... values) {
            super.onProgressUpdate(values);
        }
    }
}
```

9.4.5 具体设备页面

实现“返回”、设备名称显示、设备图片呈现、“打开”“关闭”功能。页面布局如图 9-6 所示。

“返回”功能的作用是返回上一级页面；在“返回”后面显示当前设备的名称。设备图片采用设备实物图；“打开”按键和“关闭”按键分别实现对设备的开关控制。

程序的界面布局代码清单：CH09\WareHousing\res\layout\activity_equipment.xml。

图 9-6 设备页面控制图

```
<?xml version="1.0" encoding="utf-8"?>
<RelativeLayout xmlns:android="http://schemas.
android.com/apk/res/android"
    android:layout_width="fill_parent"
    android:layout_height="fill_parent"
    android:background="@drawable/main_ default_bg">
    <include android:id="@id/bt_createtask_title_
layout"
        layout="@layout/activity_common_title_ bar" />
    <RelativeLayout
        android:id="@+id/RelativeLayout1"
        android:layout_width="match_parent"
        android:layout_height="match_parent"
```

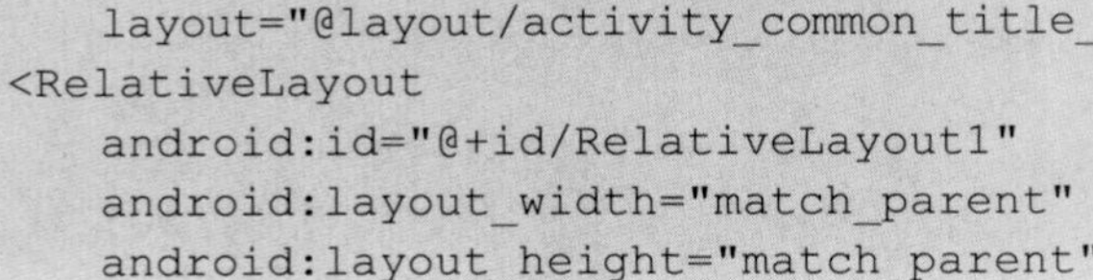

```
        android:layout_below="@id/bt_createtask_title_layout"
        android:layout_centerHorizontal="true"
        android:layout_marginLeft="15dp"
        android:layout_marginRight="15dp"
        android:background="@drawable/content_default_bg"
        android:orientation="vertical"
        android:visibility="visible" >
        <com.zigcloud.warehousing.widget.Gallery3D
            android:id="@+id/gal_rooms"
            android:layout_width="fill_parent"
            android:layout_height="wrap_content"
            android:layout_alignParentBottom="true"
            android:layout_alignParentLeft="true"
            android:spacing="30dp"
            android:unselectedAlpha="128" />
        <LinearLayout
            android:id="@+id/LinearLayout1"
            android:layout_width="match_parent"
            android:layout_height="match_parent"
            android:layout_above="@+id/gal_rooms"
            android:layout_alignParentTop="true"
            android:orientation="vertical" >
            <include
                android:id="@+id/view_loading_error"
                android:layout_width="match_parent"
                android:layout_height="match_parent"
                layout="@layout/view_loading_error"
                android:visibility="gone" />
            <include
                android:id="@+id/view_loading"
                android:layout_width="match_parent"
                android:layout_height="match_parent"
                layout="@layout/view_loading"
                android:visibility="gone" />
            <LinearLayout
              android:id="@+id/ll_equipment"
              android:layout_width="match_parent"
              android:layout_height="match_parent"
              android:orientation="vertical" >
            </LinearLayout>
        </LinearLayout>
    </RelativeLayout>
</RelativeLayout>
```

程序清单：CH09\WareHousing\src\com\zigcloud\warehousing\activity\EquipmentActivity.java。

```
package com.zigcloud.warehousing.activity;
public class EquipmentActivity extends BaseActivity{ /*** 设备* */
  private TextView titlebar_left;
  private TextView titlebar_title;
```

```
    private String nodeId;
    private String nodeName;
    private String nodeTypeId;
    private String nodeTypeName;
    private  String dataValueString;
    private String updateTimeString;
    protected void onCreate(Bundle savedInstanceState) {
        super.onCreate(savedInstanceState);
        setContentView(R.layout.activity_equipment);
        MainApplication.getInstance().addActivity(this);
        nodeId=getIntent().getStringExtra("nodeId");
        nodeName=getIntent().getStringExtra("nodeName");
        nodeTypeId=getIntent().getStringExtra("nodeTypeId");
        nodeTypeName=getIntent().getStringExtra("nodeTypeName");
        dataValueString=getIntent().getStringExtra("dataValueString");
        updateTimeString=getIntent().getStringExtra("updateTimeString");
        titlebar_left=(TextView) findViewById(R.id.titlebar_left);
        titlebar_left.setOnClickListener(new View.OnClickListener() {
            public void onClick(View v) {
                onBackPressed();
            }
        });
        titlebar_title=(TextView) findViewById(R.id.titlebar_title);
        titlebar_title.setText(nodeName);
        initalEquipment();
    }
    private void initalEquipment(){ /*** 初始化设备信息* */
        LinearLayout ll_equipment=(LinearLayout) findViewById(R.id.ll_ equipment);
        if(ll_equipment!=null){
            if(nodeTypeId!=null){
                BaseEquipmentWidget equipmentWidget=null;
                if(nodeTypeId.equals("20")){
                    equipmentWidget=new CurtainWidget(EquipmentActivity.this);
                }
                else if(nodeTypeId.equals("24")){
                    equipmentWidget=new  LampWidget(EquipmentActivity.this);
                }
                else if(nodeTypeId.equals("25")){
                    equipmentWidget=new BedLampWidget(EquipmentActivity.this);
                }
                else if(nodeTypeId.equals("26")){
                    equipmentWidget=new WallLampWidget(EquipmentActivity.this);
                }
                else{
                    equipmentWidget=new BaseEquipmentWidget(EquipmentActivity.this);
                }
                equipmentWidget.setNodeId(nodeId);
                equipmentWidget.setNodeTypeName(nodeTypeName);
```

```
                equipmentWidget.setNodeDataString(dataValueString);
                equipmentWidget.setUpdateTimeString(updateTimeString);
                equipmentWidget.addSendCmdListener(new BaseEquipmentWidget.
Listener(){
                    public void sendCmd(String nodeId, String stateFlag,String
stateValue) {
                        if(mSendCmdHttpRequestTask.getStatus()!=Status.RUNNING){
                            mSendCmdHttpRequestTask=new SendCmdHttpRequestTask();
                            mSendCmdHttpRequestTask.execute(nodeId,   stateFlag,
stateValue) ;
                        }
                    }
                });
                ll_equipment.addView(equipmentWidget,new LayoutParams(LayoutParams.
MATCH_PARENT,LayoutParams.MATCH_PARENT));
            }
        }
    }
    protected EquipmentDAO mEquipmentDAO=new EquipmentDAO();/*** 设备控制业
务类**/
    /*** 异步发送命令任务类**/
    protected  SendCmdHttpRequestTask  mSendCmdHttpRequestTask=new  Send
CmdHttpRequestTask();     /** * 切换当前模式* */
    public class SendCmdHttpRequestTask extends  AsyncTask<String, Integer,
ControlResultJson>{
        protected ControlResultJson doInBackground(String... params) {
            if(params!=null&&params.length>2){
                mEquipmentDAO.sendCmd(params[0], params[1], params[2]);
            }
            return null;
        }
    }
    protected EquipmentHttpRequestTask mEquipmentHttpRequestTask=new Equipment
HttpRequestTask();
    /*** 获取 equipment* */
    protected  class  EquipmentHttpRequestTask  extends  AsyncTask<String,
Integer, EquipmentJson>{
        protected void onPreExecute() {
            super.onPreExecute();
        }
        protected void onProgressUpdate(Integer... values) {
            super.onProgressUpdate(values);
        }
        protected EquipmentJson doInBackground(String... arg0) {
            String nodeId=arg0!=null&&arg0.length>0?arg0[0]:null;
            return mEquipmentDAO.getByNodeId(nodeId);
        }
        protected void onPostExecute(EquipmentJson result) {
```

```
            super.onPostExecute(result);
            BaseEquipmentEntity baseEquipment=BaseEquipmentEntity.parse(result);
            if(baseEquipment!=null){
            }
        }
    }
}
```

9.4.6 物品出库页面

实现物品出库功能。页面布局如图 9-7 所示。

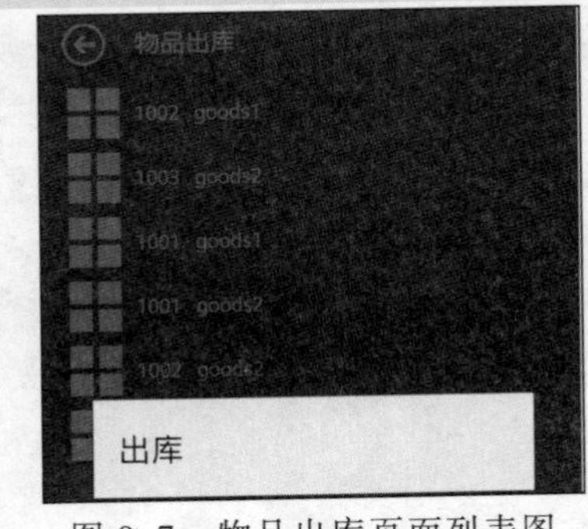

图 9-7 物品出库页面列表图

程序的界面布局代码如：CH09\WareHousing\res\layout\activity_goods_outport.xml。

```
<?xml version="1.0" encoding="utf-8"?>
<RelativeLayout xmlns:android="http://schemas.
android.com/apk/res/android"
    android:layout_width="fill_parent"
    android:layout_height="fill_parent"
    android:background="@drawable/main_ default_bg" >
    <include
        android:id="@+id/bt_createtask_title_layout"
        layout="@layout/activity_common_title_bar" />
    <RelativeLayout
        android:id="@+id/RelativeLayout1"
        android:layout_width="match_parent"
        android:layout_height="match_parent"
        android:layout_below="@id/bt_createtask_title_layout"
        android:layout_centerHorizontal="true"
        android:layout_marginLeft="15dp"
        android:layout_marginRight="15dp"
        android:background="@drawable/content_default_bg"
        android:orientation="vertical"
        android:visibility="visible" >
        <GridView
            android:id="@+id/gridView1"
            android:layout_width="match_parent"
            android:layout_height="wrap_content"
            android:layout_alignParentLeft="true"
            android:layout_alignParentTop="true"
            android:horizontalSpacing="10dp"
            android:numColumns="1"
            android:verticalSpacing="10dp"
            android:visibility="visible" >
        </GridView>
        <include
            android:id="@+id/view_loading_error"
            android:layout_width="match_parent"
            android:layout_height="match_parent"
            android:layout_alignParentLeft="true"
            android:layout_below="@+id/gdv_models"
```

```
        layout="@layout/view_loading_error"
        android:visibility="gone" />
    </RelativeLayout>
</RelativeLayout>
```

源代码清单：CH09\WareHousing\src\com\zigcloud\warehousing\activity\GoodsOutport Activity.java。

```
package com.zigcloud.warehousing.activity;
public class GoodsOutportActivity extends BaseActivity{/*** 物品出库**/
    private TextView titlebar_left;
    private TextView titlebar_title;
    private GridView goodsGridView;
    private ArrayList<HashMap<String,Object>>goodsArrayList=new ArrayList
<HashMap<String, Object>>();
    private HashMap<String,Object>goodsHashMap=new HashMap<String, Object>();
    private SimpleAdapter goodsAdapter;
    private GoodsDAO mGoodsDAO=new GoodsDAO();/*** 物品业务处理类**/
    private GoodsRequestTask mGoodsRequestTask=new GoodsRequestTask();/***
获取物品信息任务类**/
    private GoodsOutportTask mGoodsOutportTask=new GoodsOutportTask();/***
获取物品出库**/
    protected void onCreate(Bundle savedInstanceState) {
        super.onCreate(savedInstanceState);
        setContentView(R.layout.activity_goods_outport);
        MainApplication.getInstance().addActivity(this);
        titlebar_left=(TextView) findViewById(R.id.titlebar_left);
        titlebar_left.setOnClickListener(new View.OnClickListener() {
            public void onClick(View v) {
                onBackPressed();
            }
        });
        titlebar_title=(TextView) findViewById(R.id.titlebar_title);
        titlebar_title.setText(R.string.title_goods_outport);
        initialGoodsGridView();
    }
    private void initialGoodsGridView(){/*** 初始化情景模式列表**/
        goodsGridView=(GridView) findViewById(R.id.gridView1);
        goodsAdapter=new SimpleAdapter(getApplicationContext(),goodsArrayList,
R.layout.activity_goods_list_item, new String[] {"Image", "goodsCardId",
"goodsName"},new int[] {R.id.itemImage,R.id.tv_goods_cardid,R.id.tv_goods_name});
        goodsGridView.setAdapter(goodsAdapter);
        goodsGridView.setOnItemClickListener(new OnItemClickListener() {
            @SuppressWarnings({ "unchecked" })
            public void onItemClick(AdapterView<?> arg0, View arg1, int arg2,
  long arg3) {
                HashMap<String,Object>item=(HashMap<String,Object>) arg0.
getItemAtPosition(arg2);
                final String goodsCardId= item.get("goodsCardId").toString();
                AlertDialog.Builder builder=new AlertDialog.Builder(GoodsOutport
Activity.this).setIcon(R.drawable.ic_launcher).setItems(new String[]{"出库
"},new DialogInterface. OnClickListener() {
                    public void onClick(DialogInterface dialog, int which) {
```

```
                    switch(which){
                        case 0:
                            mGoodsOutportTask=new GoodsOutportTask();
                            mGoodsOutportTask.execute(goodsCardId);
                            break;
                    }
                }
            });
            builder.create().show();
        }
    });
    mGoodsRequestTask.execute();
}
/*** 获取物品信息任务类**/
private class GoodsRequestTask extends AsyncTask<String,Integer,List<GoodsJson>>{
    protected List<GoodsJson> doInBackground(String... params) {
        return mGoodsDAO.getAll();
    }
    protected void onPreExecute() {
        super.onPreExecute();
    }
    protected void onProgressUpdate(Integer... values) {
        super.onProgressUpdate(values);
    }
    protected void onPostExecute(List<GoodsJson> result) {
        super.onPostExecute(result);
        if(result!=null){
            goodsArrayList.clear();
            GoodsJson goodsEntity=null;
            for(int i=0;i<result.size();i++){
                goodsEntity= result.get(i);
                if(goodsEntity!=null){
                    goodsHashMap=new HashMap<String, Object>();
                    goodsHashMap.put("Image", R.drawable.metro_home_blacks_
scan_code);
                    goodsHashMap.put("goodsCardId", goodsEntity.cardNum);
                    goodsHashMap.put("goodsName", goodsEntity.dataName);
                    goodsHashMap.put("goodsAddress",goodsEntity.originPlace);
                    goodsHashMap.put("goodsUser", goodsEntity.staffName);
                    goodsArrayList.add(goodsHashMap);
                }
            }
            goodsAdapter.notifyDataSetChanged();
        }
        else{
            findViewById(R.id.view_loading_error).setVisibility(0);
        }
    }
}
```

```
    private class GoodsOutportTask extends AsyncTask<String, Integer,
ControlResultJson>{/*** 物品出库**/
        protected ControlResultJson doInBackground(String... params) {
            if(params!=null&&params.length>0)
                return mGoodsDAO.goodsOutport(params[0]);
            return null;
        }
        protected void onPreExecute() {
            super.onPreExecute();
        }
        protected void onProgressUpdate(Integer... values) {
            super.onProgressUpdate(values);
        }
        protected void onPostExecute(ControlResultJson result) {
            super.onPostExecute(result);
            String resStr=null;
            if(result!=null){
                resStr=result.flag==0?"出库成功！":"出库失败";
            }
            else{
                resStr="出库失败";
            }
            mGoodsRequestTask=new GoodsRequestTask();
            mGoodsRequestTask.execute();
            Toast.makeText(GoodsOutportActivity.this, resStr, Toast.LENGTH_
SHORT).show();
        }
    }
}
```

9.5 本章小结

本章通过智能仓储项目，让学生了解企业项目的整个开发流程：项目立项、需求分析、总体设计、详细设计、编码实现、系统调试。同时通过这个项目理解以前章节中学习到的知识并灵活应用到实际项目中。

参 考 文 献

[1] 杨丰盛．Android应用开发揭秘[M]．北京：机械工业出版社，2010.
[2] 李宁．Android开发权威指南[M]．北京：人民邮电出版社，2011.
[3] 盖索林．Google Android开发入门指南．2版[M]．北京：人民邮电出版社，2009.
[4] 梅尔．Android 2高级编程[M]．2版．王超，译．北京：清华大学出版社，2010.
[5] 余志龙，王世江．Google Android SDK开发范例大全．2版[M]．北京：人民邮电出版社，2010.
[6] 李宁．Android/OPhone开发完全讲义[M]．北京：水利水电出版社，2010.
[7] 李刚．疯狂Android讲义[M]．北京：电子工业出版社，2011.
[8] 汪永松．Android平台开发之旅[M]．北京：机械工业出版社，2010.
[9] E2EColud工作室．深入浅出Google Android[M]．北京：人民邮电出版社，2009.
[10] 梅尔．Android高级编程[M]．王鹏杰，霍建同，译．北京：清华大学出版社，2010.
[11] 李华忠，梁永生，刘涛．Android应用程序设计教程[M]．北京：人民邮电出版社，2013.